OFFICIAL SQA PAST PAPERS WITH ANSWERS

INTERMEDIATE 2

PHYSICS
2006-2010

First exam published in 2006.
Published by Bright Red Publishing Ltd, 6 Stafford Street, Edinburgh EH3 7AU
tel: 0131 220 5804 fax: 0131 220 6710 info@brightredpublishing.co.uk www.brightredpublishing.co.uk

ISBN 978-1-84948-129-8

A CIP Catalogue record for this book is available from the British Library.

Bright Red Publishing is grateful to the copyright holders, as credited on the final page of the book, for permission to use their material.
Every effort has been made to trace the copyright holders and to obtain their permission for the use of copyright material.
Bright Red Publishing will be happy to receive information allowing us to rectify any error or omission in future editions.

2006

[BLANK PAGE]

X069/201

NATIONAL
QUALIFICATIONS
2006

WEDNESDAY, 17 MAY
1.00 PM – 3.00 PM

PHYSICS
INTERMEDIATE 2

Read Carefully

Reference may be made to the Physics Data Booklet

1 All questions should be attempted.

Section A (questions 1 to 20)

2 Check that the answer sheet is for Physics Intermediate 2 (Section A).

3 For this section of the examination you must use an **HB pencil** and, where necessary, an eraser.

4 Check that the answer sheet you have been given has **your name**, **date of birth**, **SCN** (Scottish Candidate Number) and **Centre Name** printed on it.

Do not change any of these details.

5 If any of this information is wrong, tell the Invigilator immediately.

6 If this information is correct, **print** your name and seat number in the boxes provided.

7 There is **only one correct** answer to each question.

8 Any rough working should be done on the question paper or the rough working sheet, **not** on your answer sheet.

9 At the end of the exam, put the **answer sheet for Section A inside the front cover of your answer book**.

10 Instructions as to how to record your answers to questions 1–20 are given on page three.

Section B (questions 21 to 31)

11 Answer the questions numbered 21 to 31 in the answer book provided.

12 **All answers must be written clearly and legibly in ink**.

13 Fill in the details on the front of the answer book.

14 Enter the question number clearly in the margin of the answer book beside each of your answers to questions 21 to 31.

15 Care should be taken to give an appropriate number of significant figures in the final answers to calculations.

SCOTTISH
QUALIFICATIONS
AUTHORITY

©

DATA SHEET

Speed of light in materials

Material	Speed in m/s
Air	3.0×10^8
Carbon dioxide	3.0×10^8
Diamond	1.2×10^8
Glass	2.0×10^8
Glycerol	2.1×10^8
Water	2.3×10^8

Speed of sound in materials

Material	Speed in m/s
Aluminium	5200
Air	340
Bone	4100
Carbon dioxide	270
Glycerol	1900
Muscle	1600
Steel	5200
Tissue	1500
Water	1500

Gravitational field strengths

	Gravitational field strength on the surface in N/kg
Earth	10
Jupiter	26
Mars	4
Mercury	4
Moon	1.6
Neptune	12
Saturn	11
Sun	270
Venus	9

Specific heat capacity of materials

Material	Specific heat capacity in J/kg °C
Alcohol	2350
Aluminium	902
Copper	386
Glass	500
Ice	2100
Iron	480
Lead	128
Oil	2130
Water	4180

Specific latent heat of fusion of materials

Material	Specific latent heat of fusion in J/kg
Alcohol	0.99×10^5
Aluminium	3.95×10^5
Carbon dioxide	1.80×10^5
Copper	2.05×10^5
Iron	2.67×10^5
Lead	0.25×10^5
Water	3.34×10^5

Melting and boiling points of materials

Material	Melting point in °C	Boiling point in °C
Alcohol	−98	65
Aluminium	660	2470
Copper	1077	2567
Glycerol	18	290
Lead	328	1737
Iron	1537	2747

Specific latent heat of vaporisation of materials

Material	Specific latent heat of vaporisation in J/kg
Alcohol	11.2×10^5
Carbon dioxide	3.77×10^5
Glycerol	8.30×10^5
Turpentine	2.90×10^5
Water	22.6×10^5

Radiation weighting factors

Type of radiation	Radiation weighting factor
alpha	20
beta	1
fast neutrons	10
gamma	1
slow neutrons	3

SECTION A

For questions 1 to 20 in this section of the paper the answer to each question is either A, B, C, D or E. Decide what your answer is, then, using your pencil, put a horizontal line in the space provided—see the example below.

EXAMPLE

The energy unit measured by the electricity meter in your home is the

    A kilowatt-hour

    B ampere

    C watt

    D coulomb

    E volt.

The correct answer is **A**—kilowatt-hour. The answer **A** has been clearly marked in **pencil** with a horizontal line (see below).

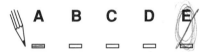

Changing an answer

If you decide to change your answer, carefully erase your first answer and, using your pencil, fill in the answer you want. The answer below has been changed to **E**.

[Turn over

SECTION A

Answer questions 1–20 on the answer sheet.

1. A car travels with an initial speed of 10 m/s. It now accelerates steadily to 30 m/s in 5 s.

   Which row shows the car's acceleration and average speed during this time?

	Acceleration (m/s²)	Average Speed (m/s)
A	2	10
B	2	20
C	4	20
D	4	30
E	8	30

2. Which of these physical quantities are equivalent?

   A Mass *and* weight

   B Mass *and* acceleration due to gravity

   C Weight *and* acceleration due to gravity

   D Weight *and* gravitational field strength

   E Acceleration due to gravity *and* gravitational field strength

3. A space vehicle of mass 120 kg is falling vertically towards a planet. The gravitational field strength at this point is 3·5 N/kg.

   The vehicle fires a rocket engine which applies a steady upward force of 660 N to the vehicle.

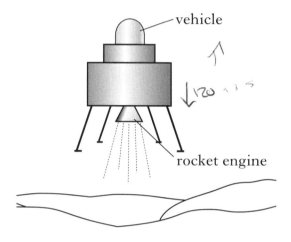

vehicle

rocket engine

   Initially the vehicle will

   A move towards the surface, accelerating

   B move towards the surface at steady speed

   C move towards the surface, decelerating

   D move away from the surface, accelerating

   E move away from the surface at steady speed.

4. A person sits on a chair which rests on the Earth. The person exerts a downward force on the chair.

Which of the following is the reaction to this force?

A   The force of the person on the Earth

B   The force of the person on the chair

C   The force of the Earth on the person

D   The force of the chair on the Earth

E   The force of the chair on the person

5. An electric motor raises a lift of mass 288 kg through a height of 15 m. The input energy to the motor is 54 000 J.

The percentage efficiency of the motor is given by $E_p = Fd$
                      = 15

A $\dfrac{288 \times 15 \times 10}{54\,000}$

B $\dfrac{54\,000 \times 10}{288 \times 15 \times 100}$

C $\dfrac{288 \times 15 \times 100}{54\,000}$

D $\dfrac{288 \times 15 \times 10 \times 100}{54\,000}$

E $\dfrac{54\,000 \times 100}{288 \times 10 \times 15}$.

6. A solid substance is placed in an insulated container and heated. The graph shows how the temperature of the substance varies with time.

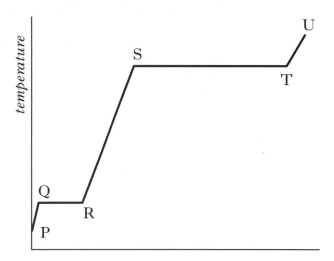

To calculate the specific latent heat of fusion of the substance, a student would use the time from section

A   PQ

B   QR

C   RS

D   ST

E   TU.

7. Which of the following statements is/are correct?

I   The voltage of a battery is the number of joules of energy it gives to each coulomb of charge.

II   A battery only has a voltage when it is connected in a complete circuit.

III   Electrons are free to move within an insulator.

A   I only

B   II only

C   III only

D   II and III only

E   I, II and III

[Turn over

8. A student suspects that ammeter A_1 may be inaccurate. Ammeter A_2 is known to be accurate.

Which of the following circuits should be used to compare A_1 with A_2?

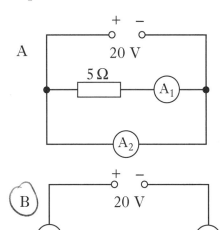

A

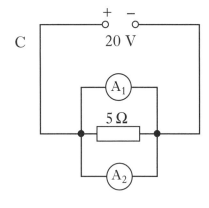

B

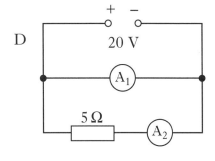

C

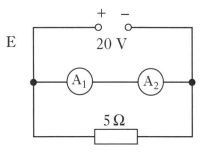

D

E

9. A circuit is set up as shown.

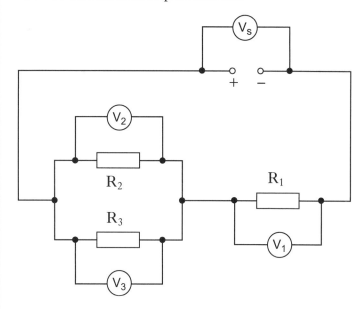

Which of the following statements about the readings on the voltmeters **must always** be true?

I $V_1 = V_2$

II $V_2 = V_3$

III $V_s = V_1 + V_2$

A   II only

B   I and II only

C   I and III only

D   II and III only

E   I, II and III

10. Three resistors are connected as shown.

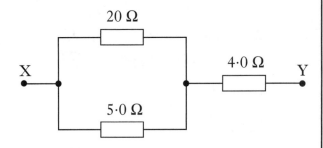

The resistance between X and Y is

A $0.5\ \Omega$

B $2.0\ \Omega$

C $4.25\ \Omega$

D $8.0\ \Omega$

E $29\ \Omega$.

11. A battery is connected in series to a lamp and resistor as shown.

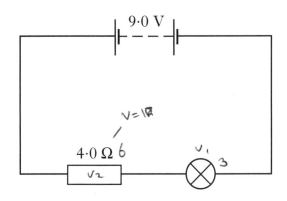

The current in the lamp is $1.5\ A$.

The power developed in the lamp is

A $3.0\ W$

B $4.5\ W$

C $6.0\ W$

D $9.0\ W$

E $13.5\ W$.

$P = VI$

$P = I^2R$

$P = VI =$

12. A student investigates the effect of moving a magnet into and out of a coil.

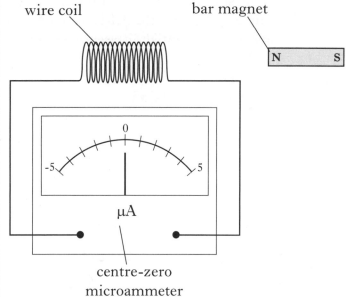

centre-zero
microammeter

Which of the following statements is/are correct?

I   When the magnet is moving into the coil, the meter shows a current.

II   The greatest current is measured when the magnet is stationary inside the coil.

III   When the magnet is moving out of the coil, the meter shows no current.

A   I only

B   II only

C   I and II only

D   II and III only

E   I, II and III

[Turn over

13. A light-emitting diode (LED) is used to show that a car windscreen heater is switched on.

The supply voltage is 14·5 V. The current through the LED is 5·0 mA, when the potential difference across it is 2·3 V.

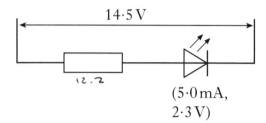

14·5 V

12.2

(5·0 mA, 2·3 V)

The resistance of the series resistor is

A   0·46 Ω

B   2·90 Ω

C   460 Ω

D   2440 Ω

E   2900 Ω.

12.2×0.05

$R = \frac{V}{I} =$

$V = IR$

$\frac{V}{IR}$

14. An amplifier has a voltage gain of 500. A signal of voltage 3·6 mV and frequency 256 Hz is applied to the input of the amplifier.

Which row shows the voltage and frequency of the signal at the amplifier output?

$V_g = \frac{out}{in}$ $V_g = \frac{out}{in}$

$out = V_g \times in$
$= 500 \times 3.6 \times 10^{-3}$

	Voltage	Frequency
A	7·2 μV	256 Hz
B	3·6 mV	128 kHz
C	1·8 V	256 Hz
D	1·8 V	128 kHz
E	1800 V	256 Hz

15. A beam of light has a wavelength of 4·80 × 10⁻⁷ m in air. The frequency of this light is

$f = \quad v = f\lambda \quad \frac{v}{\lambda}$

A   1·60 × 10⁻¹⁵ Hz $f = \frac{v}{\lambda}$

B   2·40 × 10⁻¹⁵ Hz $f = \frac{3 \cdot 00 \times 10^8}{4.80 \times 10^{-7}}$

C   7·08 × 10⁸ Hz $f =$

D   4·17 × 10¹⁴ Hz

E   6·25 × 10¹⁴ Hz.

16. Which of the following statements is/are correct?

   I   A long sighted person cannot focus sharply on near objects.

   II   In a short sighted person's eye, rays from a distant object focus behind the retina.

   III   As lens power increases, the focal length decreases.

A   I only

B   I and II only

C   I and III only

D   II and III only

E   I, II and III

17. A student is asked to write down some types of electromagnetic waves in order of increasing wavelength. The student's answer is **not** correct.

X-rays	Ultraviolet	Infrared	Visible light	Microwaves

Which **two** of these should be exchanged to make the student's answer correct?

A   X-rays and infrared

B   Visible light and infrared

C   Infrared and ultraviolet

D   Infrared and microwaves

E   X-rays and microwaves

18. Below are three statements about radiation.

    I   The half life of a radioactive source is half of the time it takes for its activity to reduce to zero.

    II The activity of a radioactive source is the number of decays per minute.

    III The risk of harm from radiation is not the same for all types of tissue.

Which statement or statements is/are true?

A   I only

B   II only

C   III only

D   II and III only

E   I, II and III

19. A worker in a nuclear power station is accidentally exposed to 3·0 mGy of gamma radiation and 0·50 mGy of fast neutrons.

The radiation weighting factor for gamma radiation is 1 and for fast neutrons is 10.

The total equivalent dose received by the worker, in mSv, is

A   3·50

B   8·00

C   11·0

D   35·0

E   38·5.

$H = D \times W_R$

$= 0.003$

$H = D \times W_R$

$= 0.005$

$H = D \times W_R$

$= 3 \times 1$

$= 3$

$= 8$

$H = D \times W_R$

$= 0.50 \times 10$

$= 5$

20. The diagram shows a nuclear reactor in a power station.

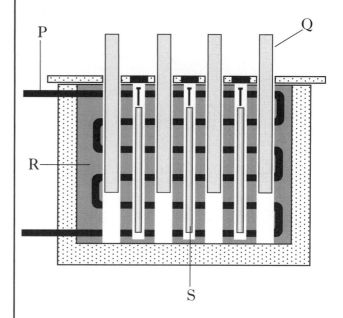

Which row shows the parts of the reactor?

	P	Q	R	S
A	moderator	fuel rod	coolant	control rod
B	control rod	moderator	coolant	fuel rod
C	moderator	control rod	coolant	fuel rod
D	coolant	control rod	moderator	fuel rod
E	coolant	fuel rod	moderator	control rod

[Turn over

$\dfrac{14}{20}$

[BLANK PAGE]

SECTION B

Marks

Write your answers to questions 21–31 in the answer book.

All answers must be written clearly and legibly in ink.

21. In a mountain-bike competition, a competitor starts from rest at the top of a hill. He pedals downhill and after 2·5 s he passes point X which is 3 m lower than the start. The total mass of the bike and competitor is 90 kg.

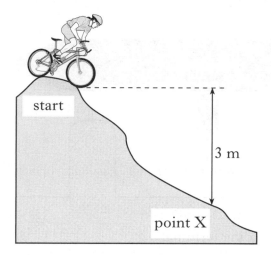

A speed time graph for this part of the competitor's journey is shown below.

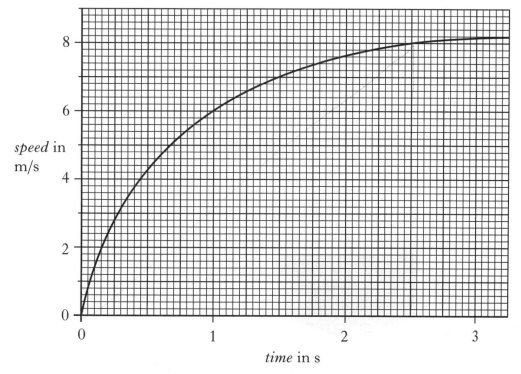

(a) Calculate the decrease in gravitational potential energy of the competitor and bike between the start and point X.   **2**

(b) Calculate the kinetic energy of the competitor and bike at point X.   **2**

(c) Explain the difference between your answers to (a) and (b).   **2**

(d) (i) What happens to the acceleration of the competitor during the first 2·5 s?   **1**

   (ii) Explain, in terms of forces, why this happens.   **1**

(8)

Marks

22. A fully laden oil tanker of mass 7.5×10^8 kg leaves a loading terminal.

Its engine and propellers produce a forward force of 6.0×10^6 N. A tugboat pushes against one side of the tanker as shown. The tug applies a pushing force of 8.0×10^6 N.

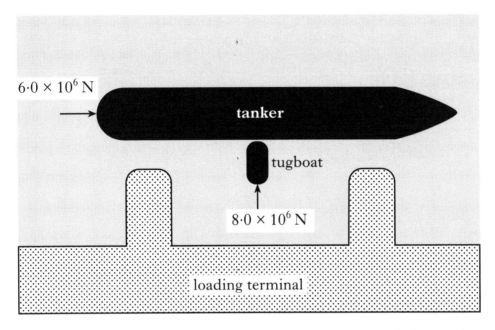

(a) Using a scale diagram or otherwise, find the size of the resultant of these two forces.   2

(b) Calculate the initial acceleration of the tanker.   2

(c) Out in the open sea, the side of the tanker is struck by a wave once every 16 s. The speed of the waves is 12·5 m/s.

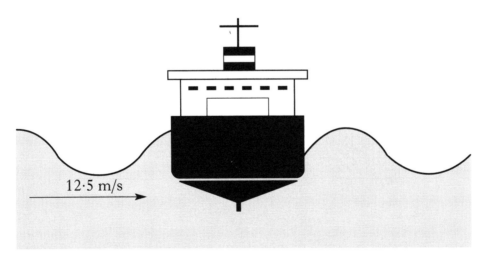

For these waves, calculate:

(i) the frequency;   1

(ii) the wavelength.   2

(7)

Marks

23. In a circus trapeze act, gymnast A has a mass of 60 kg. Gymnast B has a mass of 50 kg.

Gymnast A swings down on the trapeze and collides with gymnast B. They move off together at 4·8 m/s.

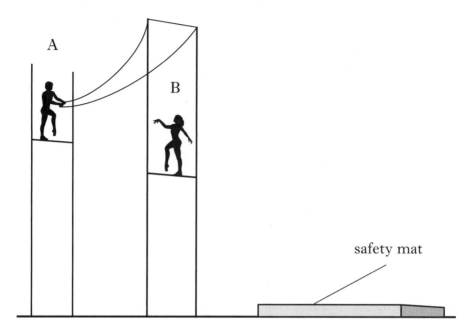

(a) Calculate the total momentum of the two gymnasts just after the collision.   2

(b) Calculate the speed of gymnast A just before the collision.   2

(c) At the point of collision, gymnast A lets go of the trapeze. At this instant, the pair are travelling horizontally. They fall together for 0·65 s until they land on a safety mat.

   (i) Calculate the horizontal distance they travel until they reach the mat.   2

   (ii) Calculate the **vertical** speed with which they strike the mat.   2

                             (8)

[Turn over

Marks

24. A blacksmith cools a hot iron horse-shoe of mass 0·75 kg by dropping it into water. The mass of the water is 15 kg and its initial temperature is 17 °C. Heat energy from the iron warms the water until both iron and water are at 23 °C.

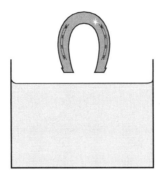

Data on page two will be required for this question.

(a) Calculate the heat energy absorbed by the water.   2

(b) Calculate the initial temperature of the horse-shoe.   3

(c) State **one** assumption required for the calculation in part (b).   1

(d) What would happen to the temperature rise of the liquid if the blacksmith had replaced the water with the same mass of oil? You **must** explain your answer.   2

(8)

Marks

25. An electric locomotive pulls a car shuttle train through the Channel Tunnel. The maximum power of the locomotive is 7·0 MW.

Current is supplied to the locomotive through an overhead wire at a voltage of 25 kV. A transformer reduces this voltage to 2·0 kV to operate the electric motors. A simplified diagram below shows the circuit.

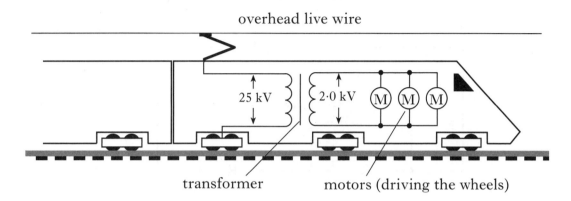

(a) The secondary coil of the transformer has 400 turns. Calculate the number of turns on the primary coil.   2

(b) Calculate the current in the secondary coil when the locomotive is at maximum power.   2

(c) Over part of the journey the train travels at a steady speed. It covers a distance of 540 m in a time of 15 s.

Calculate the maximum pulling force of the locomotive over this part of the journey.   3

(7)

[Turn over

Marks

26. (*a*) Explain the difference between direct current (d.c.) and alternating current (a.c.) in terms of the movement of charges in a conductor.

2

(*b*) A student investigates voltages in the following a.c. circuit, using two oscilloscopes connected as shown.

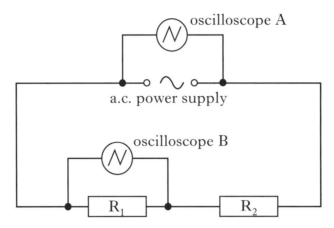

On oscilloscope A, the voltage scale is set at 5 V/cm.

On oscilloscope B, the voltage scale is set at 2 V/cm.

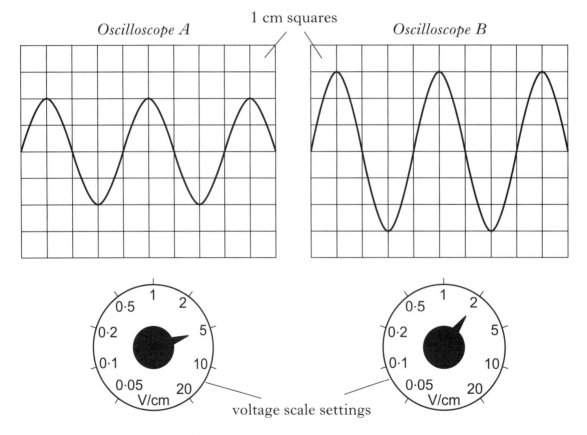

What is the peak voltage across

(i)   the power supply terminals? **1**

(ii) resistor R_1? **1**

(iii) resistor R_2? **1**

Marks

26. (continued)

(*c*) How does the quoted value of the a.c. supply voltage compare to your answer in (*b*)(i)?

1

(*d*) The student now connects some LEDs to a 2 V d.c. supply as shown.

2 V d.c.

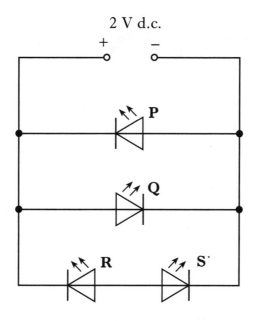

Which of the LEDs P, Q, R and S will light?

1

(*e*) The student now replaces the 2 V d.c. supply with a 2 V a.c. supply as shown.

2 V a.c.

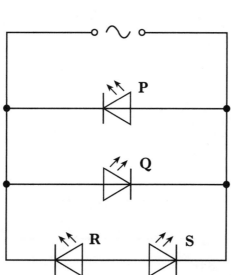

Which LED or LEDs will **now** light?

1

(8)

[Turn over

27. A student investigates solar cells connected in series. She uses a lamp with a dimmer control, a light meter, and a voltmeter as shown.

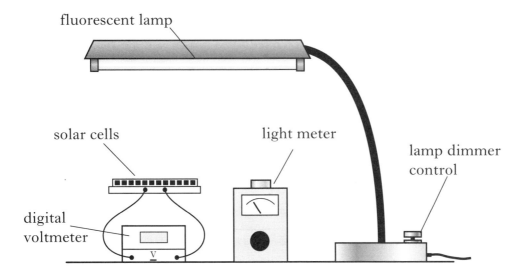

From her results, the student plots the following graph of the output voltage of the solar cells against the reading shown on the light meter.

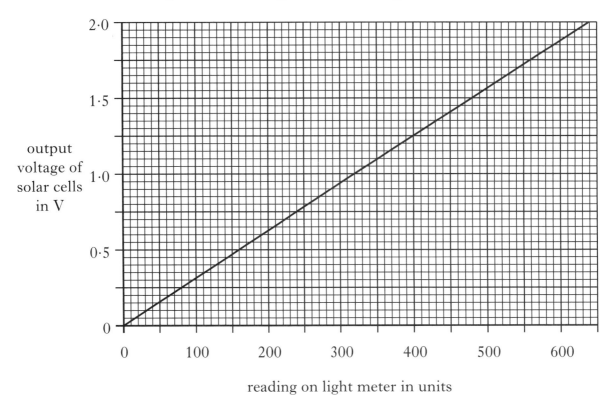

Marks

27. (continued)

(a) What reading on the light meter (in units) gives an output voltage of 0·7 V from the cells?

1

(b) Why should the cells be positioned at the same height as the light meter?

1

(c) There are four solar cells connected in series. The circuit symbol for one solar cell is

Sketch a circuit diagram of the solar cells connected to the voltmeter.

1

(d) The student now constructs the following circuit to show how the solar cells could operate the motor-driven sun shade above a shop window.

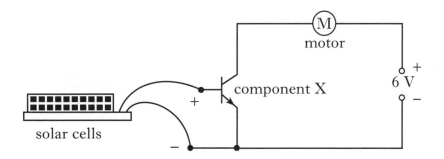

(i) Name component X.

1

(ii) Explain how this circuit operates the motor of the sun shade when the sunlight has become sufficiently bright.

3

(7)

[Turn over

Marks

28. A student uses a lens of focal length 200 mm to produce a bright, sharp image of a lamp filament on a piece of white card.

The lamp filament is positioned at a distance of 300 mm from the lens.

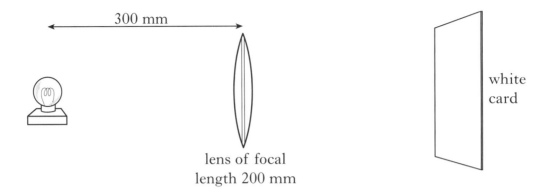

The student completes an accurate scale diagram of this experiment on the graph paper shown below. Points marked F are each one focal length from the lens. Point 2F is two focal lengths away. Two rays of light have been drawn in. The scale is shown. The piece of card and the image have not been shown.

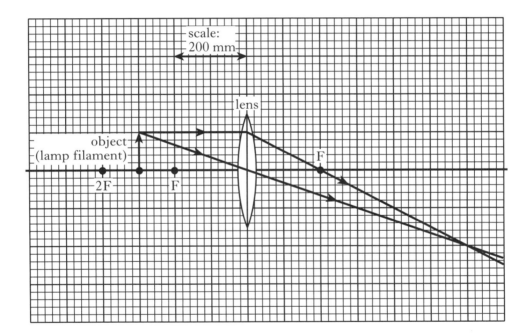

(a) By considering the scale diagram, answer the following questions.

   (i) What is the distance between the lens and the sharply focussed image on the card?   **1**

   (ii) How does the height of the image compare with the height of the object?   **1**

   (iii) State **one** other difference between the image and the object.   **1**

Marks

28.   (continued)

(*b*) The student now moves the object further away from the lens. In which direction must the card be moved to keep the image sharp?   **1**

(*c*) Calculate the power of the lens used by the student.   **2**

(*d*) A film projector in a cinema has a lens which forms an image of the film on a large, distant screen. The distance between the lens and the film is adjusted to produce a sharp image.

Describe the change which must be made to the distance between the lens and the film if the projector is moved to a smaller cinema where the screen is closer to the projector.   **1**

(7)

[Turn over

Marks

29. (*a*) A large industrial laser is used to cut metal sheets in a factory. For safety, the laser beam travels to the metal along hollow tubes with jointed "elbows". There is a plane mirror inside each "elbow" joint.

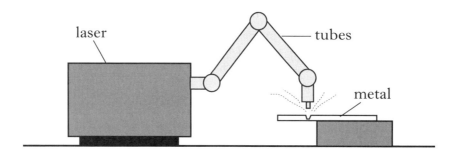

At one joint in the tube, the laser beam must change direction by 110°.

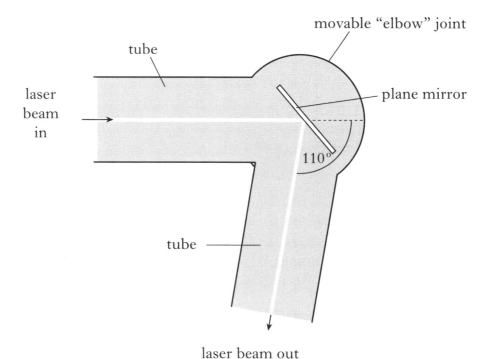

laser beam out

    (i)   What is the angle of incidence of the laser beam at the mirror?   **1**

    (ii)   What is the angle of reflection of the laser beam at the mirror?   **1**

Marks

29.   (continued)

(b) A student aims a laser beam at a triangular glass prism as shown. The beam changes direction at point X.

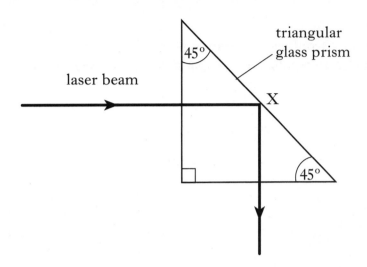

(i) Name the optical effect which occurs at point X.   1

(ii) Suggest a suitable value of the critical angle for the type of glass used for this prism.

You **must** explain your answer.   2

(5)

[Turn over

Marks

30. A spark counter consists of a thin bare wire at a high voltage, mounted on insulators beneath an earthed metal grid. There is an insulating air gap between the wire and the grid. The apparatus is connected to a sensitive microammeter and a high voltage supply as shown. The voltage of the supply is 5 kV.

When a student brings a radioactive source close to the spark counter, the air between the wire and grid is ionised and sparks jump between the wire and the grid.

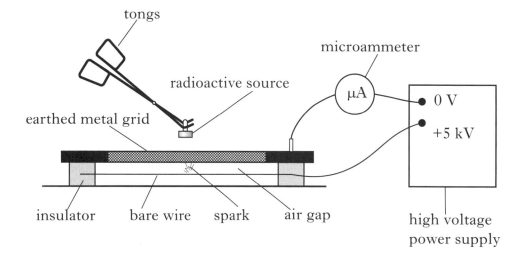

(a) The radioactive source emits alpha particles and beta particles.

State what is meant by:

    (i)   an alpha particle;   1

    (ii)   a beta particle.   1

(b) The student finds that if the source is 3 cm from the grid, there is almost continuous sparking. When the distance is increased to 6 cm, there are very few sparks.

    (i)   Explain what is meant by *ionisation*.   1

    (ii)   Which of the two types of radiation in (a) above is more effective at producing sparks? Explain your answer using the student's results.   2

(c) The student now fixes the source 5 cm above the grid. Over a period of 1 minute, the student counts 87 sparks. During this period the average reading on the microammeter is 0·29 µA.

Calculate the average quantity of charge which is transferred from the wire to the grid by each spark.   3

(8)

Marks

31. The table shows properties of some isotopes of the element iodine.

Isotope	Radiation emitted	Half-life
Iodine-127	none	–
Iodine-128	beta	25 minutes
Iodine-129	beta	16 million years
Iodine-131	beta	8·1 days
Iodine-135	beta	6·7 hours

(a) Explain what is meant by the term *half-life*.　　　　　　　1

(b) The activity of a sample of iodine-131 is 56·0 MBq.

How long will it take for its activity to reach 1·75 MBq?　　　2

(c) A patient suffers from cancer of the thyroid gland. This cancer is treated with an injection of a radioactive iodine isotope, which becomes concentrated in the thyroid gland. The thyroid receives a large dose of radiation for several hours, but surrounding tissues receive much less. Next day, when the activity of the isotope has decreased to a safe level, the patient can return home.

Which of the above isotopes would be the most suitable to treat the patient?

Explain your answer.　　　　　　　2

(d) Iodine is a necessary mineral in the diet. Some people do not receive sufficient iodine in their diet to remain healthy.

Which of the above iodine isotopes should be given to such people to supplement their diet?

Explain your answer.　　　　　　　2

　　　　　　　(7)

[END OF QUESTION PAPER]

[BLANK PAGE]

2007

[BLANK PAGE]

X069/201

NATIONAL
QUALIFICATIONS
2007

WEDNESDAY, 16 MAY
1.00 PM – 3.00 PM

PHYSICS
INTERMEDIATE 2

Read Carefully

Reference may be made to the Physics Data Booklet

1   All questions should be attempted.

Section A (questions 1 to 20)

2   Check that the answer sheet is for Physics Intermediate 2 (Section A).

3   For this section of the examination you must use an **HB pencil** and, where necessary, an eraser.

4   Check that the answer sheet you have been given has **your name**, **date of birth**, **SCN** (Scottish Candidate Number) and **Centre Name** printed on it.

   Do not change any of these details.

5   If any of this information is wrong, tell the Invigilator immediately.

6   If this information is correct, **print** your name and seat number in the boxes provided.

7   There is **only one correct** answer to each question.

8   Any rough working should be done on the question paper or the rough working sheet, **not** on your answer sheet.

9   At the end of the exam, put the **answer sheet for Section A inside the front cover of your answer book**.

10 Instructions as to how to record your answers to questions 1–20 are given on page three.

Section B (questions 21 to 31)

11 Answer the questions numbered 21 to 31 in the answer book provided.

12 **All answers must be written clearly and legibly in ink**.

13 Fill in the details on the front of the answer book.

14 Enter the question number clearly in the margin of the answer book beside each of your answers to questions 21 to 31.

15 Care should be taken to give an appropriate number of significant figures in the final answers to calculations.

SCOTTISH
QUALIFICATIONS
AUTHORITY

©

DATA SHEET

Speed of light in materials

Material	Speed in m/s
Air	$3 \cdot 0 \times 10^8$
Carbon dioxide	$3 \cdot 0 \times 10^8$
Diamond	$1 \cdot 2 \times 10^8$
Glass	$2 \cdot 0 \times 10^8$
Glycerol	$2 \cdot 1 \times 10^8$
Water	$2 \cdot 3 \times 10^8$

Speed of sound in materials

Material	Speed in m/s
Aluminium	5200
Air	340
Bone	4100
Carbon dioxide	270
Glycerol	1900
Muscle	1600
Steel	5200
Tissue	1500
Water	1500

Gravitational field strengths

	Gravitational field strength on the surface in N/kg
Earth	10
Jupiter	26
Mars	4
Mercury	4
Moon	$1 \cdot 6$
Neptune	12
Saturn	11
Sun	270
Venus	9

Specific heat capacity of materials

Material	Specific heat capacity in J/kg °C
Alcohol	2350
Aluminium	902
Copper	386
Glass	500
Ice	2100
Iron	480
Lead	128
Oil	2130
Water	4180

Specific latent heat of fusion of materials

Material	Specific latent heat of fusion in J/kg
Alcohol	$0 \cdot 99 \times 10^5$
Aluminium	$3 \cdot 95 \times 10^5$
Carbon dioxide	$1 \cdot 80 \times 10^5$
Copper	$2 \cdot 05 \times 10^5$
Iron	$2 \cdot 67 \times 10^5$
Lead	$0 \cdot 25 \times 10^5$
Water	$3 \cdot 34 \times 10^5$

Melting and boiling points of materials

Material	Melting point in °C	Boiling point in °C
Alcohol	−98	65
Aluminium	660	2470
Copper	1077	2567
Glycerol	18	290
Lead	328	1737
Iron	1537	2747

Specific latent heat of vaporisation of materials

Material	Specific latent heat of vaporisation in J/kg
Alcohol	$11 \cdot 2 \times 10^5$
Carbon dioxide	$3 \cdot 77 \times 10^5$
Glycerol	$8 \cdot 30 \times 10^5$
Turpentine	$2 \cdot 90 \times 10^5$
Water	$22 \cdot 6 \times 10^5$

Radiation weighting factors

Type of radiation	Radiation weighting factor
alpha	20
beta	1
fast neutrons	10
gamma	1
slow neutrons	3

SECTION A

For questions 1 to 20 in this section of the paper the answer to each question is either A, B, C, D or E. Decide what your answer is, then, using your pencil, put a horizontal line in the space provided—see the example below.

EXAMPLE

The energy unit measured by the electricity meter in your home is the

    A kilowatt-hour

    B ampere

    C watt

    D coulomb

    E volt.

The correct answer is **A**—kilowatt-hour. The answer **A** has been clearly marked in **pencil** with a horizontal line (see below).

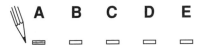

Changing an answer

If you decide to change your answer, carefully erase your first answer and, using your pencil, fill in the answer you want. The answer below has been changed to **E**.

[Turn over

SECTION A

Answer questions 1–20 on the answer sheet.

1. In the following statements X, Y and Z represent physical quantities.

   X is the displacement of an object in a given time.

   Y is the change in velocity of an object in a given time.

   Z is the distance travelled by an object in a given time.

   Which row in the table shows the quantities represented by X, Y and Z?

	X	Y	Z
A	speed	acceleration	velocity
B	velocity	speed	acceleration
C	acceleration	velocity	speed
D	acceleration	speed	velocity
E	velocity	acceleration	speed

2. Two forces act on an object as shown.

   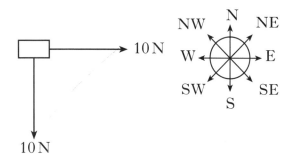

   The angle between the forces is 90°.

   The resultant force is

   A   zero
   B   14 N SE   ✓
   C   14 N NE
   ~~D   20 N SE~~
   E   20 N NE.

3. A moving vehicle X has a mass of 600 kg. It collides with a stationary vehicle Y of mass 900 kg.

   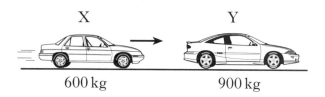

   600 kg   900 kg

   The two vehicles lock together.

   The speed of the vehicles immediately after the collision is 8·0 m/s.

   The speed of vehicle X just before the collision is

   A   8·0 m/s
   B   12 m/s
   C   13 m/s
   D   16 m/s
   E   20 m/s.   ✓

$$M_A u_A + M_B u_B = M_A v_A + M_B v_B$$

$$= 600 \times u_A + 900 \times 0 = 600 \times 8 + 900 \times 8$$

$$= 600 u_A = 12000$$

$$u_A = \frac{12000}{600}$$

$$= 20 \text{ m/s}$$

4. A ball rolls down a runway and leaves it at point R.

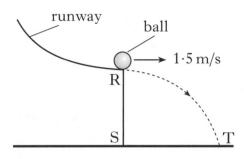

The horizontal speed of the ball at R is 1·5 m/s.

The ball takes 0·6 s to travel from R to T.

The distance ST is

A   0·40 m

B   0·90 m

C   2·5 m

D   9·0 m

E   15 m.

5. An aircraft engine exerts a force on the air.

Which of the following completes the 'Newton pair' of forces?

A   The force of the air on the aircraft engine

B   The force of friction between the aircraft engine and the air

C   The force of friction between the aircraft and the aircraft engine

D   The force of the Earth on the aircraft engine

E   The force of the aircraft engine on the Earth

6. A block of mass 6 kg is pulled across a horizontal bench by a force of 40 N as shown below.

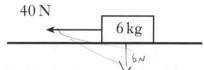

The block accelerates at 4 m/s².

The force of friction between the block and the bench is

A   zero

B   16 N

C   24 N

D   40 N

E   64 N.

7. The voltage of an electrical supply is a measure of the

A   resistance of the circuit

B   speed of the charges in the circuit

C   energy given to the charges in the circuit

D   power developed in the circuit

E   current in the circuit.

[Turn over

8. Which circuit is used to find the resistance of resistor R_2?

A

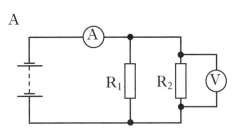

B

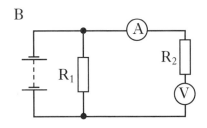

C

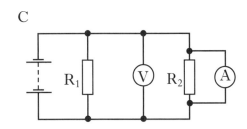

D

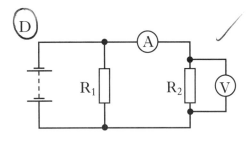

E

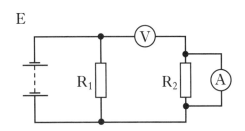

9. A circuit is set up as shown.

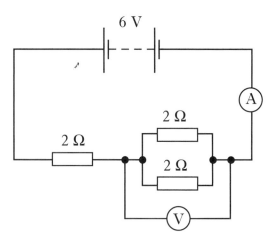

Which row in the table shows the readings on the meters?

	Reading on voltmeter (V)	Reading on ammeter (A)
A	2	1
B	2	2
C	3	2
D	4	1
E	4	2

10. An amplifier has a voltage gain of 200. A 20 mV, 100 Hz signal is applied to the input of the amplifier. Which row in the table shows the voltage and frequency of the output signal?

	Output voltage (V)	Output frequency (Hz)
A	0.1×10^{-3}	0.5
B	0.1×10^{-3}	100
C	20×10^{-3}	20 000
D	4	100
E	4	20 000

11. Which row in the table shows the symbols for an LED and an NPN transistor?

	LED	NPN transistor
A		
B		
C		
D		
E		

12. A student uses a probe connected to a meter to detect the magnetic field close to a coil of wire.

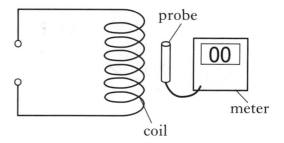

The reading on the meter is zero.

Which of the following will cause the reading on the meter to change?

A Decreasing the number of turns in the coil

B Increasing the number of turns in the coil

C Passing a current through the coil

D Replacing the coil with another coil made of thicker wire

E Replacing the coil with another coil made of thinner wire

[Turn over

13. A car headlamp is rated at 60 W.

The light produced is 20% of the total energy transferred by the lamp.

The energy transferred as light in 10 s is

A   12 J

B   120 J   ✓

C   600 J

D   3000 J

E   12 000 J.

$E = Pt$

$= 600 \div 100$

$= 6 \times 20$

14. A circuit is set up as shown.

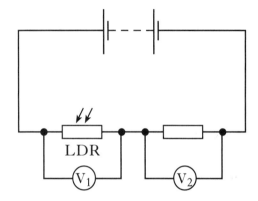

$V = IR$
$= 1 \times 2.5$

1×2

$= 2$

$V = IR$
$= 1 \times 1.5$

The initial reading on both voltmeters V_1 and V_2 is 2·5 V.

The light shining on the LDR is made brighter.

Which row in the table shows possible new readings on voltmeters V_1 and V_2?

	Reading on V_1 (V)	Reading on V_2 (V)
A	2·0	3·0
B	2·5	2·0
C	2·5	2·5
D	2·5	3·0
E	3·0	2·0

15. The diagram below shows a ray of red light entering a block of perspex.

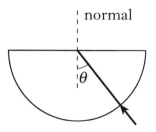

The angle θ is greater than the critical angle in the perspex for this light.

Which of the following diagrams shows the path of the ray of red light after striking the straight surface of the perspex block?

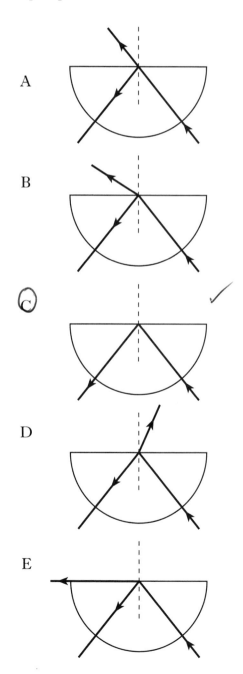

A

B

C   ✓

D

E

16. The diagram shows two rays of light incident on a curved reflector. The focal point, F, of the reflector is shown.

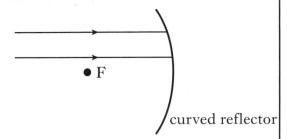

Which of the following diagrams shows the paths of the rays of light after reflection?

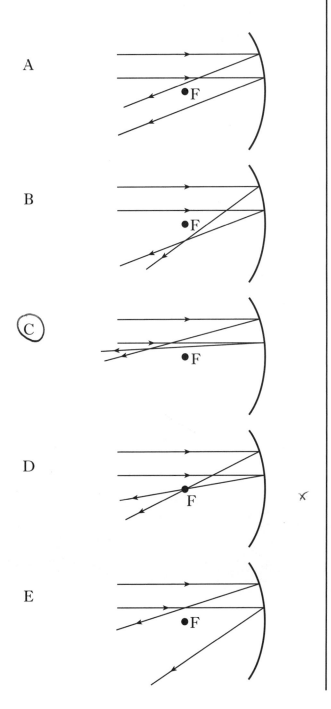

17. A student makes the following statements about members of the electromagnetic spectrum.

    I Gamma rays have a longer wavelength than X-rays.

    II Ultraviolet rays have a longer wavelength than infrared rays.

    III TV and radio waves have a longer wavelength than microwaves.

Which of the statements is/are correct?

A I only

B II only

C III only

D I and II only

E II and III only

18. Which of the following statements is/are true about fission?

    I A large nucleus is split into two smaller nuclei.

    II Two smaller nuclei join together to form a larger nucleus.

    III Fission can result in a chain reaction.

A I only

B II only

C III only

D I and III only

E II and III only

[Turn over

19. A radioactive source emits α, β and γ radiation.

Sheets of aluminium and paper are placed close to the source as shown.

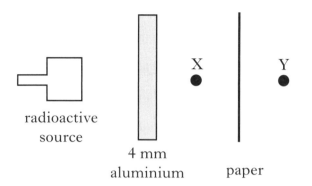

radioactive
source

4 mm
aluminium   paper

Which row in the table shows the radiation(s) from the source detected at points X and Y?

	Radiation(s) detected at X	Radiation detected at Y
A	α, γ	γ
B	β, γ	α
C	α	β
D	β	γ
E	γ	γ

−1

20. Which of the following statements about the function of parts of a nuclear reactor is/are correct?

  I The coolant removes heat from the core of the reactor.

  II Control rods contain the fuel used for the nuclear reaction.

  III The moderator slows down neutrons.

A   I only

B   I and II only

C   I and III only

D   II and III only

E   I, II and III

−1

13/17

SECTION B

Marks

Write your answers to questions 21–31 in the answer book.

All answers must be written clearly and legibly in ink.

21. A climber of mass 60 kg is attached by a rope to point A on a rock face. She climbs up to point B in 20 seconds. Point B is 3·2 m vertically above point A.

 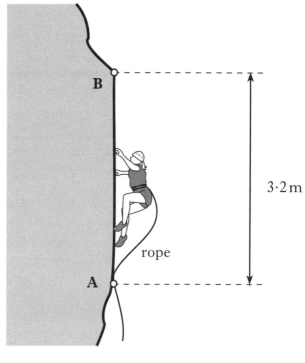

   (*a*)   (i)   Calculate the average speed of the climber between A and B.   2

         (ii)   Calculate the weight of the climber.   2

         (iii)   Calculate her gain in potential energy.   2

   (*b*)   She then loses her footing and free falls from point B. After passing point A she is held safely by the rope.

         (i)   Calculate her speed as she passes point A.   2

         (ii)   How would her actual speed when passing point A compare with the speed calculated in (*b*) (i)?

              You **must** explain your answer.   2

                                    (10)

[Turn over

Marks

22. A cyclist rides along a road.

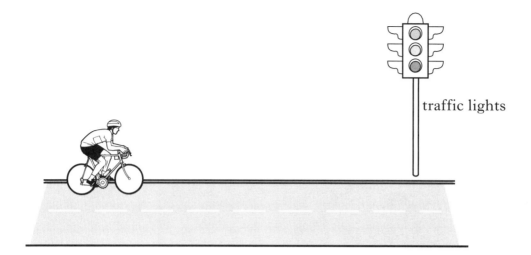

traffic lights

(a) Describe a method by which the average speed of the cyclist could be measured.

Your description must include the following:

- Measurements made
- Equipment used
- Any necessary calculations.   **3**

(b) The cyclist approaches traffic lights at a speed of 8 m/s. He sees the traffic lights turn red and 3 s later he applies the brakes. He comes to rest in a further 2·5 s.

(i) Calculate the acceleration of the cyclist whilst braking.   **2**

(ii) Sketch a speed time graph showing the motion of the cyclist from the moment the lights turn red until he stops at the traffic lights. Numerical values **must** be included.   **2**

(iii) Calculate the total distance the cyclist travels from the moment the lights turn red until he stops at the traffic lights.   **2**

(9)

Marks

23. A steam wallpaper stripper is used on the walls of a room.

Water is heated until it boils and produces steam. The plate is held against the wall and steam is released from the plate.

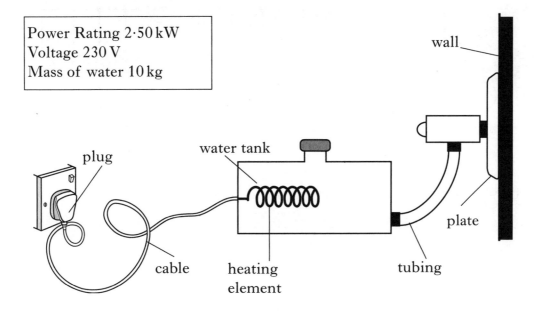

Power Rating 2·50 kW
Voltage 230 V
Mass of water 10 kg

The tank is filled with water. The water has an initial temperature of 20 °C.

(*a*) (i) Calculate the energy required to bring the water to its boiling point. 2

  (ii) Calculate the time taken for this to happen. 2

  (iii) The actual time taken for this to happen was found to be longer than that calculated in (*a*) (ii). Explain why. 1

(*b*) Calculate the current required by the wallpaper stripper. 2

(*c*) After using the wallpaper stripper for some time, 1·2 kg of water is converted into steam. Calculate the energy used to do this. 2

(9)

[Turn over

Marks

24. The following circuit shows a method of transmitting power over long distances.

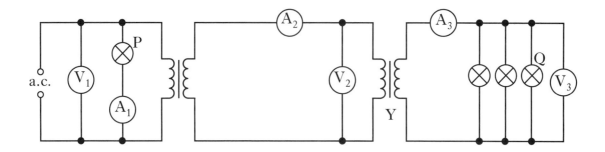

The table shows the readings on some of the meters.

V_1	6 V
A_1	60 mA
V_3	4·8 V
A_3	150 mA

(a) Why must a.c. be used in this circuit?   1

(b) All lamps are identical. Calculate the power of lamp Q.   3

(c) Transformer Y has 1000 turns in the primary coil and 50 turns in the secondary coil. Calculate the reading on voltmeter V_2 assuming the transformer is 100% efficient.   2

(d) Is the reading on A_2 larger, smaller or the same as the reading on A_3?
Explain your answer.   2

    (8)

Marks

25. A group of students visit a Laser Game Centre. The laser gun emits both a visible beam and an IR beam. Each target jacket contains three IR sensors.

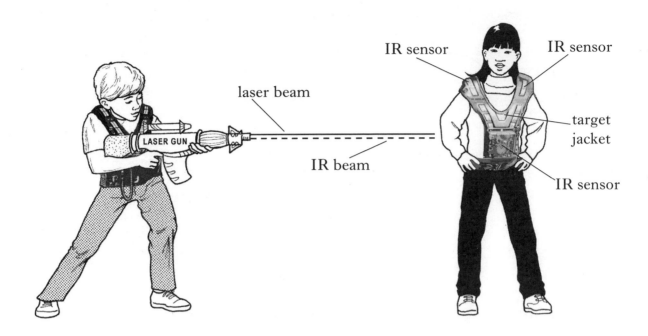

(a) (i) What does the term IR stand for? — 1

(ii) Which of the two beams arrives at the target first?

You **must** explain your answer. — 2

(b) The laser gun operates from a 7·2 V rechargeable battery. The battery is charged from the mains and takes two hours to fully recharge. A current of 3 A is used in the charging circuit.

Calculate how much charge the battery stores when fully charged. — 2

(c) When the IR beam hits a sensor on the target jacket, the following circuit is completed and the LED lights. The LED has an operating voltage of 2 V and an operating current of 15 mA. The circuit has an 8 V supply.

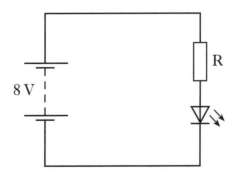

Calculate the value of resistor R. — 3

(8)

[Turn over

Marks

26. At a beauty salon, a beautician uses hot wax to help remove hair from a client's leg. It is very important that the wax remains at a constant temperature in the heating tank.

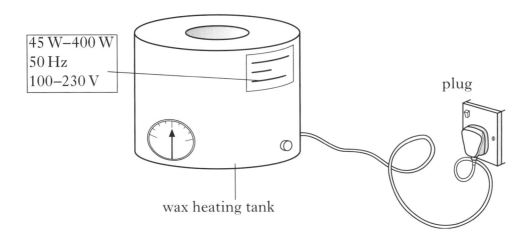

When the wax drops below a certain temperature, a heater is automatically switched on. A simplified circuit is shown.

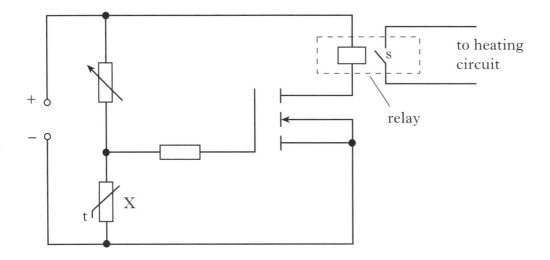

(*a*) Name component X. 1

(*b*) Explain how the circuit works to switch the heater on. 3

(*c*) What is the purpose of the variable resistor? 1

(5)

Marks

27. When the sun shines during a shower of rain, a rainbow can sometimes be seen.

The diagram shows what happens to sunlight when it enters a raindrop.

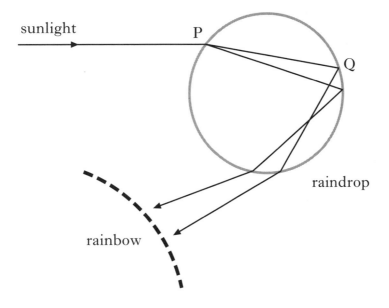

   (*a*)   (i) Name the wave effect that happens at point P.   **1**

          (ii) Name the wave effect that happens at point Q.   **1**

        (iii) Which colour of the rainbow has the longest wavelength?   **1**

   (*b*) As a raindrop falls it reaches a steady speed.

        Using Newton's laws of motion, explain why it falls at a steady speed.   **2**

                                                                         (5)

[Turn over

Marks

28. (*a*) Two types of waveform are shown.

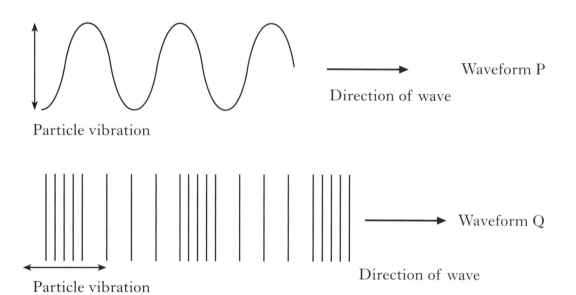

(i) Which waveform represents a longitudinal wave?   **1**

(ii) Which waveform represents a sound wave?   **1**

(*b*) A signal generator is connected to a loudspeaker which produces a sound wave of frequency 2 kHz.

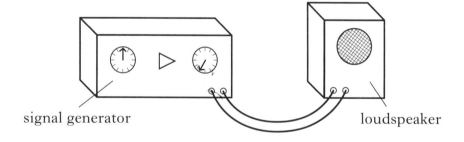

(i) Calculate the wavelength of the sound wave in air.   **2**

(ii) The loudspeaker is placed a distance of 10·2 m from a wall. Calculate the time taken for the sound to return to the loudspeaker.   **2**

(*c*) The loudspeaker is now placed in a tank of carbon dioxide gas. The frequency remains at 2 kHz.

What effect does this have on the wavelength of the sound?

Explain your answer.   **2**

(8)

Marks

29. A football player injures his leg while playing in a match.

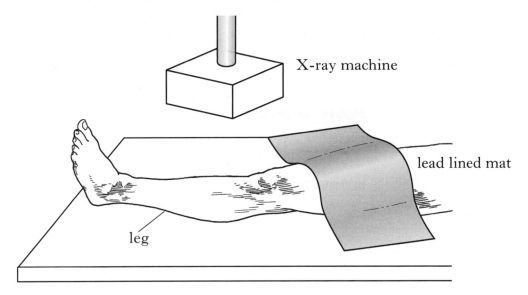

In hospital the player has three X-rays, each producing an absorbed dose of 50 μGy.

(a) The mass of the player's leg is 6 kg. Calculate the energy absorbed by the leg from the X-rays.

2

(b) Why is the rest of the player's leg covered with a lead lined mat?

1

(c) Apart from absorbed dose, name **one** other factor that contributes to biological harm.

1

(4)

[Turn over

Marks

30. A simplified diagram of a smoke detector is shown. Radiation from the source causes ionisation of the air molecules between the plates. This produces a small current in a circuit. When smoke particles pass between the plates, the current decreases and a buzzer sounds.

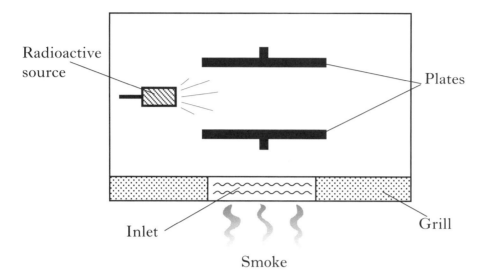

(a) (i) What is meant by *ionisation*?   1

  (ii) Should the source be an alpha, beta or gamma emitter?

Explain your answer in terms of ionisation.   2

  (iii) A manufacturer is choosing a new source for its smoke detectors. From the following information, select the most suitable source to use.

Explain your answer.   2

Source	Half-life (years)	Range (metres)
W	1	0·05
X	10	2·0
Y	100	0·05
Z	1000	2·0

(b) The smoke detector circuit contains a 9 V battery. When there is no smoke present the operating current in the circuit is 30 mA.

  (i) Calculate the resistance of the circuit.   2

  (ii) State the energy change which takes place in the buzzer.   1

                                                                                                  (8)

Marks

31. An experiment is carried out in a laboratory to determine the half-life of a radioactive source. A Geiger-Müller tube and counter are used to measure the background radiation over a period of 10 seconds. This is repeated several times and an average value of 4 counts in 10 seconds is recorded.

The apparatus shown is used to measure the count rate over a period of time. The readings are corrected for background radiation.

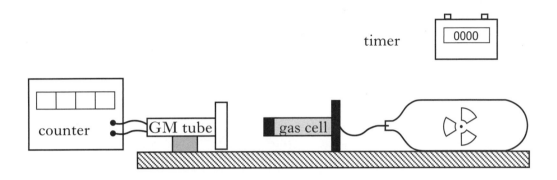

Time (minutes)	Corrected count rate
0	168
2	120
4	84
6	60
8	42
10	30
12	21

(a) Name **two** factors that affect the background count rate. 2

(b) Calculate the activity of the background radiation. 2

(c) Calculate the half-life of the radioactive source. 2

(6)

[END OF QUESTION PAPER]

[BLANK PAGE]

2008

[BLANK PAGE]

X069/201

NATIONAL
QUALIFICATIONS
2008

FRIDAY, 23 MAY
1.00 PM – 3.00 PM

PHYSICS
INTERMEDIATE 2

Read Carefully

Reference may be made to the Physics Data Booklet

1 All questions should be attempted.

Section A (questions 1 to 20)

2 Check that the answer sheet is for Physics Intermediate 2 (Section A).

3 For this section of the examination you must use an **HB pencil** and, where necessary, an eraser.

4 Check that the answer sheet you have been given has **your name**, **date of birth**, **SCN** (Scottish Candidate Number) and **Centre Name** printed on it.

   Do not change any of these details.

5 If any of this information is wrong, tell the Invigilator immediately.

6 If this information is correct, **print** your name and seat number in the boxes provided.

7 There is **only one correct** answer to each question.

8 Any rough working should be done on the question paper or the rough working sheet, **not** on your answer sheet.

9 At the end of the exam, put the **answer sheet for Section A inside the front cover of your answer book**.

10 Instructions as to how to record your answers to questions 1–20 are given on page three.

Section B (questions 21 to 31)

11 Answer the questions numbered 21 to 31 in the answer book provided.

12 **All answers must be written clearly and legibly in ink**.

13 Fill in the details on the front of the answer book.

14 Enter the question number clearly in the margin of the answer book beside each of your answers to questions 21 to 31.

15 Care should be taken to give an appropriate number of significant figures in the final answers to calculations.

DATA SHEET

Speed of light in materials

Material	Speed in m/s
Air	3.0×10^8
Carbon dioxide	3.0×10^8
Diamond	1.2×10^8
Glass	2.0×10^8
Glycerol	2.1×10^8
Water	2.3×10^8

Speed of sound in materials

Material	Speed in m/s
Aluminium	5200
Air	340
Bone	4100
Carbon dioxide	270
Glycerol	1900
Muscle	1600
Steel	5200
Tissue	1500
Water	1500

Gravitational field strengths

	Gravitational field strength on the surface in N/kg
Earth	10
Jupiter	26
Mars	4
Mercury	4
Moon	1.6
Neptune	12
Saturn	11
Sun	270
Venus	9

Specific heat capacity of materials

Material	Specific heat capacity in J/kg °C
Alcohol	2350
Aluminium	902
Copper	386
Glass	500
Ice	2100
Iron	480
Lead	128
Oil	2130
Water	4180

Specific latent heat of fusion of materials

Material	Specific latent heat of fusion in J/kg
Alcohol	0.99×10^5
Aluminium	3.95×10^5
Carbon dioxide	1.80×10^5
Copper	2.05×10^5
Iron	2.67×10^5
Lead	0.25×10^5
Water	3.34×10^5

Melting and boiling points of materials

Material	Melting point in °C	Boiling point in °C
Alcohol	−98	65
Aluminium	660	2470
Copper	1077	2567
Glycerol	18	290
Lead	328	1737
Iron	1537	2747

Specific latent heat of vaporisation of materials

Material	Specific latent heat of vaporisation in J/kg
Alcohol	11.2×10^5
Carbon dioxide	3.77×10^5
Glycerol	8.30×10^5
Turpentine	2.90×10^5
Water	22.6×10^5

Radiation weighting factors

Type of radiation	Radiation weighting factor
alpha	20
beta	1
fast neutrons	10
gamma	1
slow neutrons	3

SECTION A

For questions 1 to 20 in this section of the paper the answer to each question is either A, B, C, D or E. Decide what your answer is, then, using your pencil, put a horizontal line in the space provided—see the example below.

EXAMPLE

The energy unit measured by the electricity meter in your home is the

    A kilowatt-hour

    B ampere

    C watt

    D coulomb

    E volt.

The correct answer is **A**—kilowatt-hour. The answer **A** has been clearly marked in **pencil** with a horizontal line (see below).

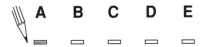

Changing an answer

If you decide to change your answer, carefully erase your first answer and, using your pencil, fill in the answer you want. The answer below has been changed to **E**.

[Turn over

SECTION A

Answer questions 1–20 on the answer sheet.

1. Which of the following is a vector quantity?

   A   Distance

   B   Energy

   C   Speed

   D   Time

   E   Velocity

2. A student walks from X to Y and then from Y to Z.

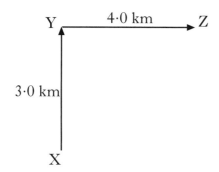

   The complete walk takes 2 hours.

   Which row in the table shows the average speed and the average velocity for the complete walk?

	Average speed	Average velocity
A	2·5 km/h	2·5 km/h at 053
B	2·5 km/h at 053	2·5 km/h
C	3·5 km/h	2·5 km/h at 053
D	3·5 km/h at 053	3·5 km/h
E	3·5 km/h	3·5 km/h at 053

3. A car travelling in a straight line decelerates uniformly from 20 m/s to 12 m/s in 4 seconds. The displacement of the car in this time is

   A   32 m

   B   48 m

   C   64 m

   D   80 m

   E   128 m.

4. An unbalanced force of one newton will make a

   A   0·1 kg mass accelerate at 1 m/s^2

   B   1 kg mass accelerate at 1 m/s^2

   C   1 kg mass accelerate at 10 m/s^2

   D   0·1 kg mass move at a constant speed of 1 m/s

   E   1 kg mass move at a constant speed of 10 m/s.

5. A trolley of mass 0·6 kg is travelling at 5 m/s along a smooth, level track.

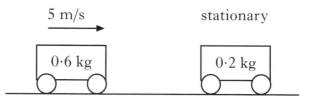

   The trolley collides with a stationary trolley of mass 0·2 kg.

   The magnitude of the total momentum of the trolleys immediately after collision is

   A   0 kg m/s

   B   1·0 kg m/s

   C   2·0 kg m/s

   D   3·0 kg m/s

   E   4·0 kg m/s.

6. A power station has an efficiency of 40%. The input power to the station is 1600 MW.

   What is the useful output power?

   A   40 MW

   B   640 MW

   C   960 MW

   D   4000 MW

   E   64000 MW

7. A sample of water is at a temperature of 100 °C. The sample absorbs $2 \cdot 3 \times 10^4$ J of energy.

   The specific latent heat of vaporisation of water is $22 \cdot 6 \times 10^5$ J/kg.

   The mass of water changed into steam at 100 °C is

   A   0·01 kg

   B   5·3 kg

   C   100 kg

   D $2 \cdot 3 \times 10^4$ kg

   E $2 \cdot 3 \times 10^6$ kg.

8. Three circuit symbols X, Y and Z are shown.

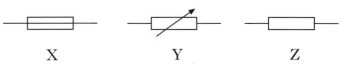

   X   Y   Z

   Which row in the table identifies the symbols X, Y and Z?

	X	Y	Z
A	thermistor	transistor	resistor
B	fuse	variable resistor	thermistor
C	transistor	fuse	variable resistor
D	fuse	variable resistor	resistor
E	variable resistor	resistor	fuse

9. A circuit is set up as shown.

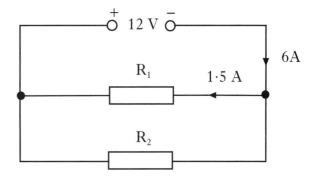

   The current from the supply is 6 A. The current in resistor R_1 is 1·5 A.

   Which row in the table shows the potential difference across resistor R_2 and the current in resistor R_2?

	Potential difference across R_2 (V)	Current in R_2 (A)
A	12	1·5
B	6	1·5
C	12	4·5
D	6	4·5
E	12	7·5

[Turn over

10. A circuit is set up as shown.

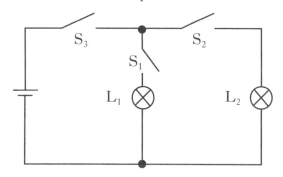

Which switch or switches must be closed to light lamp L_1 **only**?

A S_1 only

B S_2 only

C S_1 and S_2 only

D S_1 and S_3 only

E S_2 and S_3 only

11. The information shown is for an electric food mixer.

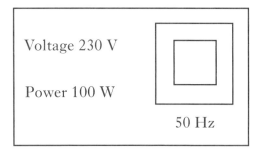

The resistance of the mixer is

A $0.43\ \Omega$

B $2.3\ \Omega$

C $4.6\ \Omega$

D $529\ \Omega$

E $23\ 000\ \Omega$.

12. When a magnet is pushed into or pulled out of a coil of wire, a voltage is induced across the ends of the coil.

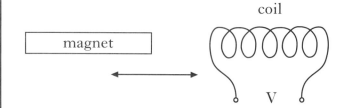

Which of the following produces the greatest induced voltage?

	Strength of magnet	Speed of magnet	Number of turns in a coil
A	weak	slow	20
B	weak	fast	40
C	strong	slow	20
D	strong	fast	20
E	strong	fast	40

13. A manufacturer states that an amplifier has a voltage gain of 15. This means that

A   the output frequency is 15 times the input frequency

B   the input frequency is 15 times the output frequency

C   the output voltage is 15 times the input voltage

D   the input voltage is 15 times the output voltage

E   the input voltage is 15 V.

14. The following diagram shows a wave.

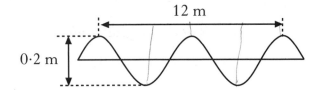

Which row in the table gives the wavelength and amplitude of the wave?

	Wavelength (m)	Amplitude (m)
A	4	0·2
B	6	0·1
C	6	0·2
D	12	0·1
E	12	0·2

15. A ray of light passes from air into a glass block as shown.

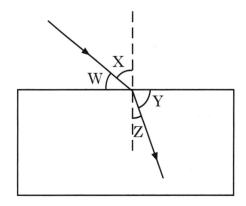

Which row in the table shows the angle of incidence and the angle of refraction?

	Angle of incidence	Angle of refraction
A	W	Z
B	W	Y
C	X	Z
D	X	Y
E	Z	X

16. A student wears glasses fitted with concave lenses. Which of the following statements is/are correct?

  I The student is short sighted.

  II Concave lenses are converging lenses.

  III The glasses help the student to see near objects clearly.

A I only

B II only

C III only

D I and II only

E I, II and III

[Turn over

17. Which row in the table describes an alpha particle, a beta particle and a gamma ray?

	Alpha particle	Beta particle	Gamma ray
A	neutron	helium nucleus	electromagnetic radiation
B	helium nucleus	electron	electromagnetic radiation
C	hydrogen nucleus	electromagnetic radiation	electron
D	helium nucleus	electromagnetic radiation	neutron
E	hydrogen nucleus	electron	electromagnetic radiation

18. For a particular radioactive source, 1800 atoms decay in a time of 3 minutes. The **activity** of this source is

A   10 Bq

B   600 Bq

C   1800 Bq

D   5400 Bq

E   324 000 Bq.

19. One gray is equal to

A   one becquerel per kilogram

B   one sievert per second

C   one joule per second

D   one sievert per kilogram

E   one joule per kilogram.

20. A student makes the following statements about nuclear reactors.

  I Fission takes place in the fuel rods.

  II The material in the control rods slows down neutrons.

  III The material in the moderator absorbs neutrons.

Which of the statements is/are correct?

A   I only

B   I and II only

C   I and III only

D   II and III only

E   I, II and III

SECTION B

Marks

Write your answers to questions 21–31 in the answer book.

All answers must be written clearly and legibly in ink.

21. Athletes in a race are recorded by a TV camera which runs on rails beside the track.

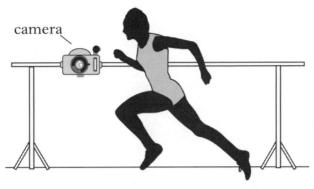

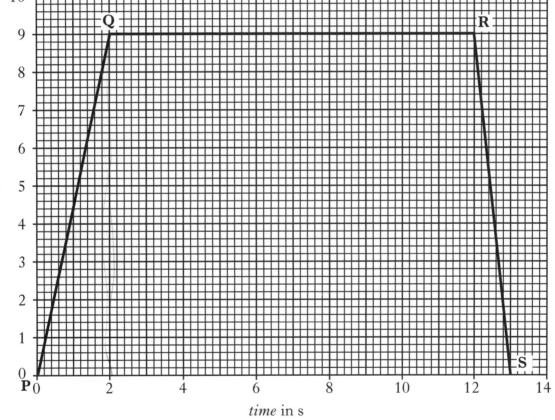

speed in m/s

time in s

The graph shows the speed of the camera during the race.

(a) Calculate the acceleration of the camera between **P** and **Q**. $4.5 \, m/s^2$ 2

(b) The mass of the camera is 15 kg.

Calculate the unbalanced force needed to produce the acceleration between **P** and **Q**. $F = ma$ $F = M \quad + \quad ma$ 2

(c) How far does the camera travel in the 13 s? $a = \frac{F}{m}$ 2

(d) The camera lens has a focal length of 200 mm.

Calculate the power of the lens. $= \frac{F}{15}$

$= 13F$

2

$F = 135 \, N$

(8)

Marks

22. A fairground ride uses a giant catapult to launch people upwards using elastic cords.

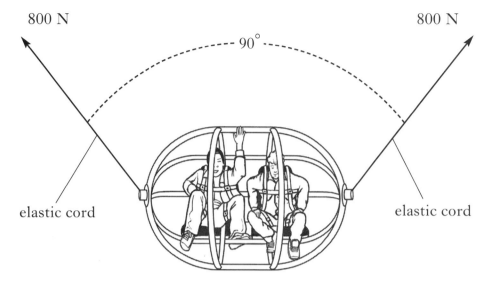

800 N   800 N

90°

elastic cord   elastic cord

(a) Each cord applies a force of 800 N and the cords are at 90° as shown. Using a scale diagram, or otherwise, find the size of the resultant of these two forces.   2

(b) The cage is now pulled further down before release. The cords provide an upward resultant force of 2700 N. The cage and its occupants have a total mass of 180 kg.

  (i) Calculate the weight of the cage and occupants.   2

  (ii) Calculate the acceleration of the cage and occupants when released.   3

(7)

Marks

23. One type of exercise machine is shown below.

(*a*) A person using this machine pedals against friction forces applied to the wheel by the brake.

A friction force of 300 N is applied at the edge of the wheel, which has a circumference of 1·5 m.

(i) How much work is done by friction in one turn of the wheel?   **2**

(ii) The person turns the wheel 500 times in 5 minutes.

Calculate the average power produced.   **3**

(*b*) The wheel is a solid aluminium disc of mass 12·0 kg.

(i) All the work done by friction is converted to heat in the disc.

Calculate the temperature rise after 500 turns.   **2**

(ii) Explain why the actual temperature rise of the disc is less than calculated in (*b*) (i).   **1**

(8)

[Turn over

Marks

24. An early method of crash testing involved a car rolling down a slope and colliding with a wall.

In one test, a car of mass 750 kg starts at the top of a 7·2 m high slope.

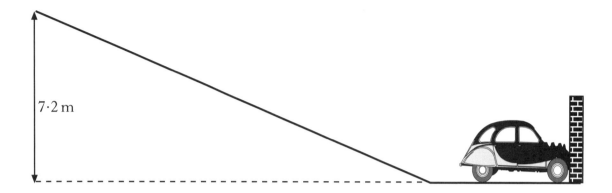

(a) Calculate the gravitional potential energy of the car at the top of the slope.   **2**

(b) (i) State the value of the kinetic energy of the car at the bottom of the slope, assuming no energy losses.   **1**

(ii) Calculate the speed of the car at the bottom of the slope, before hitting the wall.   **2**

(5)

Marks

25. Some resistors are labelled with a power rating as well as their resistance value. This is the maximum power at which they can operate without overheating.

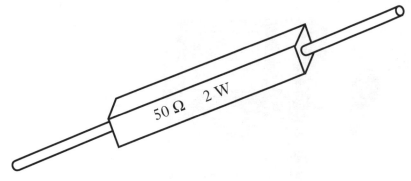

(*a*) A resistor is labelled 50 Ω, 2 W.

Calculate the maximum operating current for this resistor. 2

(*b*) Two resistors, each rated at 2 W, are connected in parallel to a 9 V d.c. supply.

They have resistances of 60 Ω and 30 Ω.

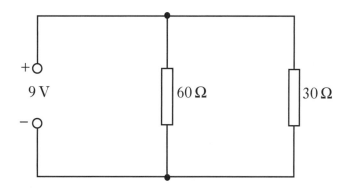

   (i) Calculate the total resistance of the circuit. 2

  (ii) Calculate the power produced in each resistor. 3

 (iii) State which, if any, of the resistors will overheat. 1

(*c*) The 9 V **d.c.** supply is replaced by a 9 V **a.c.** supply.

What effect, if any, would this have on your answers to part (*b*) (ii)? 1

(9)

[Turn over

Marks

26. A karaoke machine consists of a microphone, amplifier, loudspeaker, DVD player and screen.

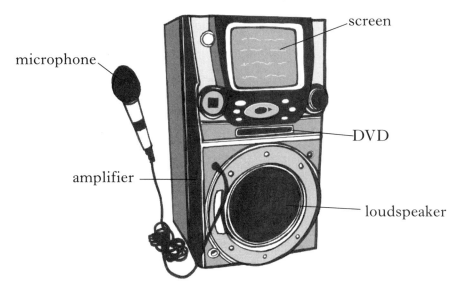

   (a) What energy change takes place in the microphone?   1

   (b) The amplifier processes the signal from the microphone.

   What effect does the amplifier have on the signal's

   (i) frequency;   1

   (ii) amplitude?   1

   (c) A singer produces a note of frequency 850 Hz. The speed of sound in air is 340 m/s.

   Calculate the wavelength of this note in air.   2

   (d) The DVD player contains a laser.

   Light from this laser enters a small glass prism as shown.

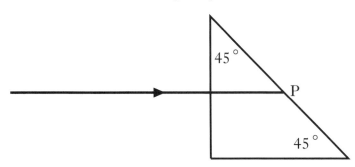

   The glass has a critical angle of 40°.

   (i) Explain what is meant by the term "critical angle".   1

   (ii) Copy and complete the diagram to show the path of the ray after it strikes point P.   2

   (8)

Marks

27. An office has an automatic window blind that closes when the light level outside gets too high.

The electronic circuit that operates the motor to close the blind is shown.

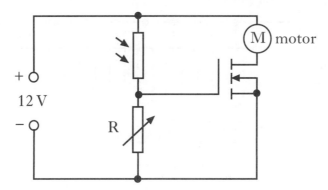

(a) The MOSFET switches on when the voltage across variable resistor R reaches 2·4 V.

    (i) Explain how this circuit works to close the blind.   3

    (ii) What is the purpose of the variable resistor R?   1

(b) The graph shows how the resistance of the LDR varies with light level.

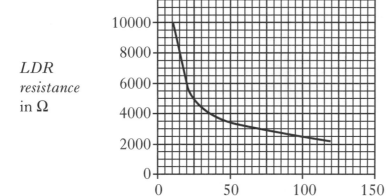

    (i) What is the resistance of the LDR when the light level is 70 units?   1

    (ii) R has a value of 600 Ω. Calculate the voltage across R when the light level is 70 units.   2

    (iii) State whether or not the blinds will close when the light level is 70 units.

        Justify your answer.   2

(9)

[Turn over

Marks

28. The rear light of a car is made up of a row of 10 **identical** red LEDs. Each LED requires 2 V and 20 mA to operate correctly.

(*a*) The circuit for this is shown.

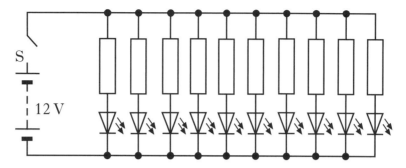

    (i)   Why does each LED need a resistor in series?   **1**

    (ii) The voltage of the car battery is 12 V.

         Calculate the value of each resistor.   **3**

    (iii) Calculate the total current, **in amperes**, from the battery when the rear light is operating correctly.   **2**

(*b*) Some car headlights require 84 V to operate. Electronic circuits are needed to convert the car battery voltage.

Part of the circuit contains a transformer as shown.

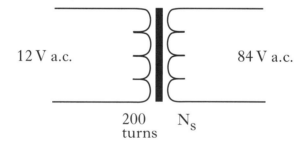

The primary coil of the transformer has 200 turns.

Calculate the number of turns, N_S, in the secondary coil.   **2**

    (8)

Marks

29. A "bug viewer" has a plastic chamber with a lens in the lid. It is used to get a magnified view of small insects placed on the base of the chamber.

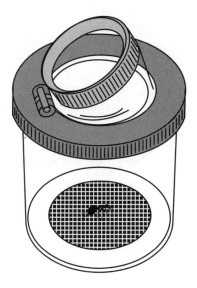

(a) What type of lens should be used? 1

(b) The lens used has a focal length of 60 mm and the base of the chamber is 30 mm from the lens.

Copy and complete this diagram by adding rays to show where the image of the bug will be formed. 2

Use the squared ruled paper provided (small squares side).

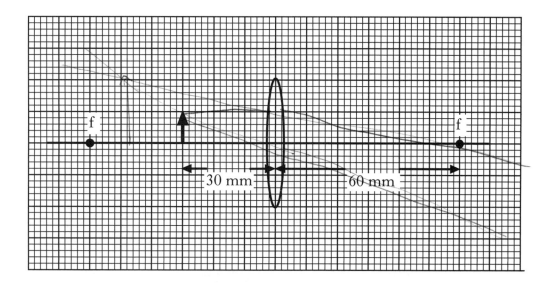

(c) How would the shape of this lens have to be altered to give it a longer focal length? 1

(d) Name the eye defect which this type of lens could correct. 1

(5)

[Turn over

Marks

30. When welders join thick steel plates it is important that the joint is completely filled with metal. This ensures there are no air pockets in the metal weld, as this would weaken the joint.

One method of checking for air pockets is to use a radioactive source on one side of the joint. A detector placed as shown measures the count rate on the other side.

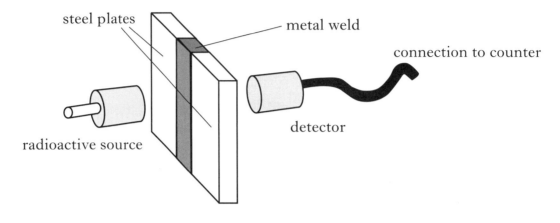

View from above

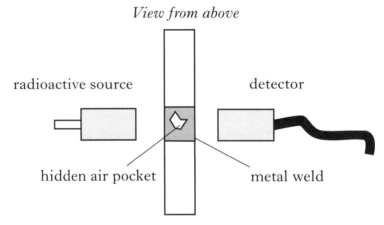

(a) The radioactive source and detector are moved along the weld. How would the count rate change when the detector moves over an air pocket?

Explain your answer. **2**

(b) Which of the radiations alpha, beta or gamma must be used?

Explain your answer. **2**

(c) X-rays are sometimes used to detect air pockets.

How does the wavelength of X-rays compare with gamma rays? **1**

(5)

Marks

31. Gold-198 is a radioactive source that is used to trace factory waste which may cause river pollution.

A small quantity of the radioactive gold is added into the waste as it enters the river. Scanning the river using radiation detectors allows scientists to trace where the waste has travelled.

Gold-198 has a half-life of 2·7 days.

(a) What is meant by the term "half-life"?   1

(b) A sample of Gold-198 has an activity of 64 kBq when first obtained by the scientists.

Calculate the activity after 13·5 days.   2

(c) Describe two precautions taken by the scientists to reduce the equivalent dose they receive while using radioactive sources.   2

(d) A scientist receives an absorbed dose of 10 mGy of alpha radiation.

(i) Calculate the equivalent dose received.   2

(ii) The risk of biological harm from radiation exposure depends on the absorbed dose and the type of radiation. Which other factor affects the risk of biological harm?   1

(8)

[END OF QUESTION PAPER]

[BLANK PAGE]

2009

[BLANK PAGE]

X069/201

| NATIONAL QUALIFICATIONS 2009 | TUESDAY, 26 MAY 1.00 PM – 3.00 PM | PHYSICS INTERMEDIATE 2 |

Read Carefully

Reference may be made to the Physics Data Booklet

1 All questions should be attempted.

Section A (questions 1 to 20)

2 Check that the answer sheet is for Physics Intermediate 2 (Section A).

3 For this section of the examination you must use an **HB pencil** and, where necessary, an eraser.

4 Check that the answer sheet you have been given has **your name**, **date of birth**, **SCN** (Scottish Candidate Number) and **Centre Name** printed on it.

Do not change any of these details.

5 If any of this information is wrong, tell the Invigilator immediately.

6 If this information is correct, **print** your name and seat number in the boxes provided.

7 There is **only one correct** answer to each question.

8 Any rough working should be done on the question paper or the rough working sheet, **not** on your answer sheet.

9 At the end of the exam, put the **answer sheet for Section A inside the front cover of your answer book**.

10 Instructions as to how to record your answers to questions 1–20 are given on page three.

Section B (questions 21 to 29)

11 Answer the questions numbered 21 to 29 in the answer book provided.

12 **All answers must be written clearly and legibly in ink**.

13 Fill in the details on the front of the answer book.

14 Enter the question number clearly in the margin of the answer book beside each of your answers to questions 21 to 29.

15 Care should be taken to give an appropriate number of significant figures in the final answers to calculations.

DATA SHEET

Speed of light in materials

Material	Speed in m/s
Air	$3 \cdot 0 \times 10^8$
Carbon dioxide	$3 \cdot 0 \times 10^8$
Diamond	$1 \cdot 2 \times 10^8$
Glass	$2 \cdot 0 \times 10^8$
Glycerol	$2 \cdot 1 \times 10^8$
Water	$2 \cdot 3 \times 10^8$

Speed of sound in materials

Material	Speed in m/s
Aluminium	5200
Air	340
Bone	4100
Carbon dioxide	270
Glycerol	1900
Muscle	1600
Steel	5200
Tissue	1500
Water	1500

Gravitational field strengths

	Gravitational field strength on the surface in N/kg
Earth	10
Jupiter	26
Mars	4
Mercury	4
Moon	$1 \cdot 6$
Neptune	12
Saturn	11
Sun	270
Venus	9

Specific heat capacity of materials

Material	Specific heat capacity in J/kg °C
Alcohol	2350
Aluminium	902
Copper	386
Glass	500
Ice	2100
Iron	480
Lead	128
Oil	2130
Water	4180

Specific latent heat of fusion of materials

Material	Specific latent heat of fusion in J/kg
Alcohol	$0 \cdot 99 \times 10^5$
Aluminium	$3 \cdot 95 \times 10^5$
Carbon dioxide	$1 \cdot 80 \times 10^5$
Copper	$2 \cdot 05 \times 10^5$
Iron	$2 \cdot 67 \times 10^5$
Lead	$0 \cdot 25 \times 10^5$
Water	$3 \cdot 34 \times 10^5$

Melting and boiling points of materials

Material	Melting point in °C	Boiling point in °C
Alcohol	−98	65
Aluminium	660	2470
Copper	1077	2567
Glycerol	18	290
Lead	328	1737
Iron	1537	2747

Specific latent heat of vaporisation of materials

Material	Specific latent heat of vaporisation in J/kg
Alcohol	$11 \cdot 2 \times 10^5$
Carbon dioxide	$3 \cdot 77 \times 10^5$
Glycerol	$8 \cdot 30 \times 10^5$
Turpentine	$2 \cdot 90 \times 10^5$
Water	$22 \cdot 6 \times 10^5$

Radiation weighting factors

Type of radiation	Radiation weighting factor
alpha	20
beta	1
fast neutrons	10
gamma	1
slow neutrons	3

SECTION A

For questions 1 to 20 in this section of the paper the answer to each question is either A, B, C, D or E. Decide what your answer is, then, using your pencil, put a horizontal line in the space provided—see the example below.

EXAMPLE

The energy unit measured by the electricity meter in your home is the

   A kilowatt-hour

   B ampere

   C watt

   D coulomb

   E volt.

The correct answer is **A**—kilowatt-hour. The answer **A** has been clearly marked in **pencil** with a horizontal line (see below).

Changing an answer

If you decide to change your answer, carefully erase your first answer and, using your pencil, fill in the answer you want. The answer below has been changed to **E**.

[Turn over

SECTION A

Answer questions 1–20 on the answer sheet.

1. Which of the following quantities requires both magnitude and direction?

   A   Mass

   B   Distance

   C   Momentum

   D   Speed

   E   Time

2. A cross country runner travels 2·1 km North then 1·5 km East. The total time taken is 20 minutes.

   The average speed of the runner is

   A   0·18 m/s

   B   2·2 m/s

   C   3·0 m/s

   D   130 m/s

   E   180 m/s.

3. The graph shows how the velocity of an object varies with time.

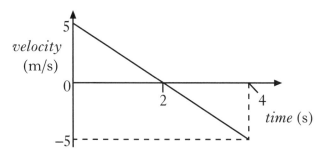

   Which row in the table shows the displacement after 4 s and the acceleration of the object during the first 4 s?

	Displacement (m)	Acceleration (m/s^2)
A	10	−10
B	10	2·5
C	0	2·5
D	0	−10
E	0	−2·5

4 A ball is thrown horizontally from a cliff as shown.

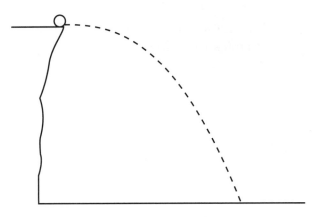

The effect of air resistance is negligible.

A student makes the following statements about the ball.

   I The vertical speed of the ball increases as it falls.

  II The vertical acceleration of the ball increases as it falls.

 III The vertical force on the ball increases as it falls.

Which of the statements is/are correct?

A   I only

B   II only

C   I and II only

D   II and III only

E   I, II and III

5. Which block has the largest resultant force acting on it?

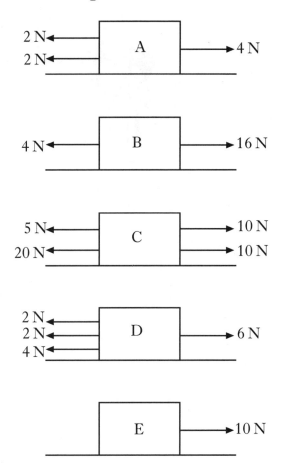

[Turn over

6. An arrow is fired from a bow as shown.

An archer pulls the string back a distance of 0·50 m. The string exerts an average force of 300 N on the arrow as it is fired. The mass of the arrow is 0·15 kg.

The maximum kinetic energy gained by the arrow is

A   23 J

B   150 J

C   600 J

D   2000 J

E   6750 J.

7. A solid substance is placed in an insulated container and is heated at a constant rate. The graph shows how the temperature of the substance changes with time.

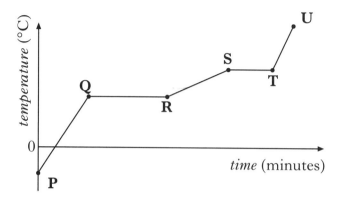

During the time interval QR, which of the following statements is/are correct?

  I There is a change in the state of the substance.

 II The substance changes state from a liquid to a gas.

III Heat is absorbed by the substance.

A   I only

B   III only

C   I and II only

D   I and III only

E   I, II and III

8. A student writes the following statements about electrical conductors.

    I   Only protons are free to move.

    II Only electrons are free to move.

    III Only negative charges are free to move.

    Which of the statements is/are correct?

    A   I only

    B   II only

    C   III only

    D   I and II only

    E   II and III only

9. A charge of 15 C passes through a resistor in 12 s. The potential difference across the resistor is 6 V.

    The power developed by the resistor is

    A   4·8 W

    B   7·5 W

    C   9·4 W

    D   30 W

    E   1080 W.

10. A circuit is set up as shown.

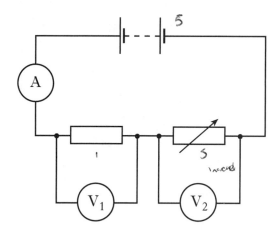

The resistance of the variable resistor is increased.

Which row in the table shows the effect on the readings on the ammeter and voltmeters?

	Reading on ammeter	Reading on voltmeter V_1	Reading on voltmeter V_2
A	decreases	decreases	decreases
B	increases	unchanged	increases
C	decreases	increases	decreases
D	increases	unchanged	decreases
E	decreases	decreases	increases

[Turn over

11. A circuit is set up as shown.

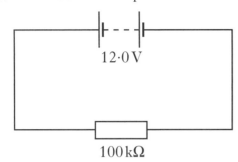

12·0 V

100 kΩ

The power supplied to the resistor is

A $1·20 \times 10^{-4}$ W

B $1·44 \times 10^{-3}$ W

C $1·44$ W

D 694 W

E $1·20 \times 10^{6}$ W.

12. Which of the following devices transforms light energy into electrical energy?

A   LED

B   Thermocouple

C   Microphone

D   Solar cell

E   Transistor

13. Which of the following is the correct symbol for an n-channel enhancement MOSFET?

A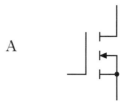

B

C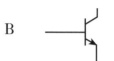

D

E

14. Which of the following is an example of a longitudinal wave?

    A   Light wave

    B   Infra-red wave

    C   Radio wave

    D   Sound wave

    E   Water wave

15. The diagram shows a list of the members of the electromagnetic spectrum in order of increasing wavelength.

gamma rays	P	ultraviolet	Q	infrared	R	TV and Radio

Which row in the table shows the radiation represented by the letters **P**, **Q** and **R**?

	P	Q	R
A	microwaves	visible light	x-rays
B	visible light	microwaves	x-rays
C	x-rays	visible light	microwaves
D	visible light	x-rays	microwaves
E	x-rays	microwaves	visible light

16. The diagram shows what happens to a ray of light when it strikes a glass block.

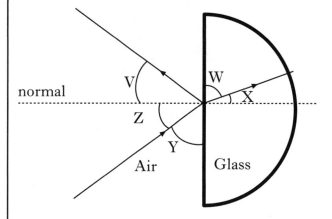

Which row in the table identifies the angle of incidence and the angle of refraction?

	Angle of Incidence	Angle of Refraction
A	V	W
B	Y	W
C	Y	X
D	Z	W
E	Z	X

[Turn over

17. The diagram below shows a simple model of an atom.

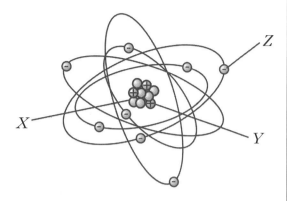

Which row in the table identifies particles X, Y and Z?

	X	Y	Z
A	electron	proton	neutron
B	proton	neutron	electron
C	neutron	electron	proton
D	electron	neutron	proton
E	neutron	proton	electron

18. A student makes the following statements about ionising radiations.

I Ionisation occurs when an atom loses an electron. ✓

II Gamma radiation produces greater ionisation (density) than alpha particles. ✗

III An alpha particle consists of 2 protons, 2 neutrons and 2 electrons.

Which of the statements is/are correct?

A I only

B II only

C I and II only

D II and III only

E I, II and III

19. A sample of tissue has a mass of $0.05\,kg$.

The tissue is exposed to radiation and absorbs $0.1\,J$ of energy in 2 minutes.

The absorbed dose is

A $0.005\,Gy$

B $0.1\,Gy$

C $0.5\,Gy$

D $2\,Gy$

E $6\,Gy$.

20. During fission, a neutron splits a uranium nucleus into two nuclei, X and Y, as shown below.

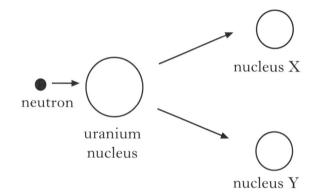

For a chain reaction to occur which of the following **must** also be released?

A   Protons

B   Electrons

C   Neutrons

D   Alpha particles

E   Gamma radiation

Candidates are reminded that the answer sheet for Section A MUST be placed INSIDE the front cover of the answer book.

SECTION B

Marks

Write your answers to questions 21–29 in the answer book.

All answers must be written clearly and legibly in ink.

21. A ski lift with a gondola of mass 2000 kg travels to a height of 540 m from the base station to a station at the top of the mountain.

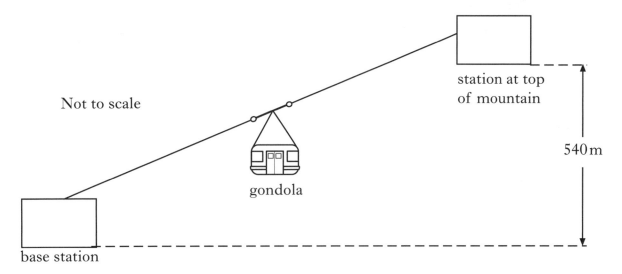

Not to scale

station at top
of mountain

540 m

gondola

base station

(a) Calculate the gain in gravitational potential energy of the gondola.   2

(b) During the journey, the kinetic energy of the gondola is 64 000 J.

Calculate the speed of the gondola.   2

(c) The ski lift requires a motor which operates at 380 V to take the gondola up the mountain. The maximum power produced is 45·6 kW.

(i) Calculate the maximum current in the motor.   2

(ii) Calculate the electrical energy used by the motor when it has been operating at its maximum power for a total time of 1 hour.   2

    (8)

[Turn over

Marks

22. A child sledges down a hill.

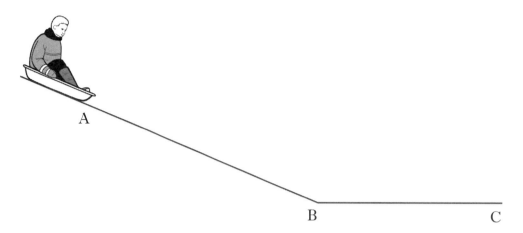

The sledge and child are released from rest at point A. They reach a speed of 3 m/s at point B.

(a) The sledge and child take 5 s to reach point B.

Calculate the acceleration.   **2**

(b) The sledge and child have a combined mass of 40 kg.

Calculate the unbalanced force acting on them.   **2**

(c) After the sledge and child pass point B, they slow down, coming to a halt at point C.

Explain this motion in terms of forces.   **2**

(6)

Marks

23. The following apparatus is used to determine the speed of a pellet as it leaves an air rifle. The air rifle fires a pellet into the plasticine, causing the vehicle to move.

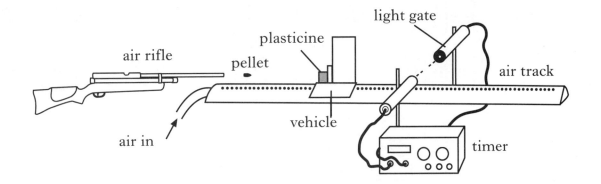

(a) Describe how the apparatus is used to determine the speed of the vehicle.

Your description must include:

- the measurements made
- any necessary calculations.   **2**

(b) The speed of the vehicle is calculated as 0·35 m/s after impact.

The mass of the pellet is $5·0 \times 10^{-4}$ kg. The mass of the vehicle and plasticine before impact is 0·30 kg.

(i) Show that the momentum of the pellet **before** impact with the plasticine is 0·105 kg m/s.   **1**

(ii) Hence, calculate the velocity of the pellet **before** impact with the plasticine.   **1**

(c) At a firing range a pellet is fired horizontally at a target 40 m away. It takes 0·20 s to reach the target.

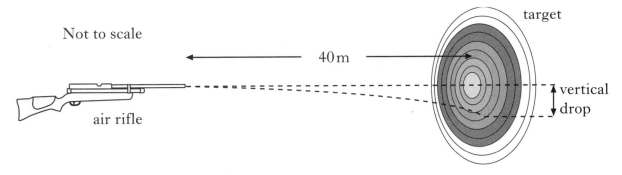

(i) Calculate the **vertical** velocity of the pellet on reaching the target.   **2**

(ii) Calculate the vertical drop.   **2**

(8)

[Turn over

Marks

24. A fridge/freezer has water and ice dispensers as shown.

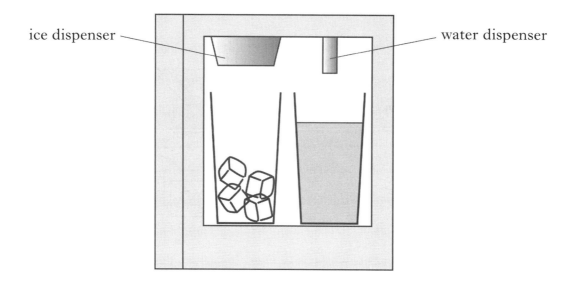

ice dispenser ⟶   ⟵ water dispenser

(a) Water of mass 0·1 kg flows into the freezer at 15 °C and is cooled to 0 °C. Calculate the energy removed when the water cools.   2

(b) Calculate how much energy is released when 0·1 kg of water at 0 °C changes to 0·1 kg of ice at 0 °C.   2

(c) The fridge/freezer system removes heat energy at a rate of 125 J/s.

   (i) Calculate the minimum time taken to produce 0·1 kg of ice from 0·1 kg of water at 15 °C.   3

   (ii) Explain why the actual time taken to make the ice will be longer than the value calculated in part (i).   2

(9)

Marks

25. A student sets up the following circuit to investigate the resistance of resistor R.

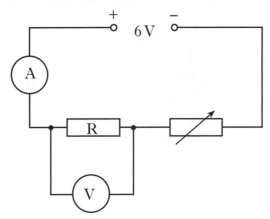

The variable resistor is adjusted and the voltmeter and ammeter readings are noted. The following graph is obtained from the experimental results.

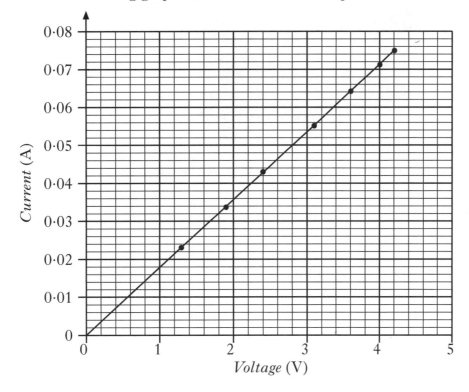

(*a*) (i) Calculate the value of the resistor R when the reading on the voltmeter is 4·2 V. **3**

(ii) Using information from the graph, state whether the resistance of the resistor R, **increases**, **stays the same** or **decreases** as the voltage increases.

Justify your answer. **2**

(*b*) The student is given a task to combine two resistors from a pack containing one each of 33 Ω, 56 Ω, 82 Ω, 150 Ω, 270 Ω, 390 Ω.

Show by calculation which **two** resistors should be used to give:

(i) the largest combined resistance; **2**

(ii) the smallest combined resistance. **2**

(9)

Marks

26. An MP3 player is charged from the mains supply of 230 V using a transformer, which has an output voltage of 5 V and an output current of 1 A.

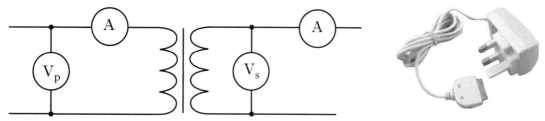

circuit diagram of transformer   MP3 Charger

(a) Calculate the current in the primary circuit.   2

(b) The MP3 player is then put on a docking station with external speakers.

docking station
with speakers   MP3 player

    (i) Calculate the resistance of a 10 W speaker when the voltage across it is 9 V.   2

    (ii) Calculate the gain of the amplifier in the docking station when the input voltage is 1·5 V.   2

(c) The input power to the amplifier is 25 W. The output power is 20 W. Calculate the efficiency of the amplifier.   2

(8)

Marks

27. A student is short sighted.

(a) (i) What does the term "short sighted" mean? 1

(ii) What type of lens is required to correct this eye defect? 1

(iii) The focal length of the lens needed to correct the student's short sight is 180 mm. Calculate the power of this lens. 2

(b) In the eye, refraction of light occurs at both the cornea and the lens. Some eye defects can be corrected using a laser. Light from the laser is used to change the shape of the cornea.

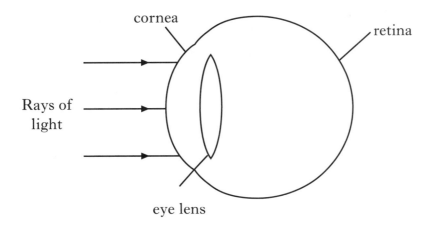

(i) State what is meant by refraction of light. 1

(ii) The laser emits light of wavelength 7×10^{-7} m.

Calculate the frequency of the light. 2

(c) Lasers can be used in optical fibres for medical purposes.

(i) Copy and complete the path of the laser light along the optical fibre. 2

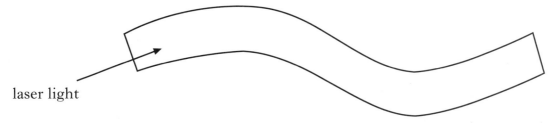

(ii) Name the effect when the laser light hits the inside surface of the fibre. 1

(10)

[Turn over

Marks

28. Parking sensors are fitted to the rear bumper of some cars. A buzzer emits audible beeps, which become more frequent as the car moves closer to an object.

emitters and sensors

Ultrasonic pulses are emitted from the rear of the car. Objects behind the car reflect the pulses, which are detected by sensors. Ultrasonic pulses travel at the speed of sound.

(a) The time between these pulses being sent and received is 2×10^{-3} s.

Calculate the distance between the object and the rear of the car. **3**

(b) At a certain distance, the buzzer beeps every 0·125 s.

Calculate the frequency of the beeps. **2**

(c) The sensor operates at a voltage of 12 V and has a current range of 20–200 mA.

Calculate the maximum power rating of the sensor. **3**

(d) An LED system can be added so that it flashes at the same frequency as the beeps from the buzzer. The LED circuit is shown below.

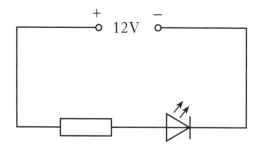

(i) A resistor is connected in series with the LED.

State the purpose of the resistor. **1**

(ii) When lit, the LED has a voltage of 3·5 V across it and a current of 200 mA.

Calculate the value of the resistor. **3**

(12)

Marks

29. A radioactivity kit includes three radioactive sources each made up as shown.

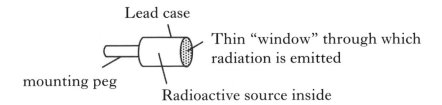

Lead case

Thin "window" through which radiation is emitted

mounting peg

Radioactive source inside

Information about these sources is given in the table below.

	Radiation Emitted	*Radioactive Element*
Source 1	Alpha	Americium 241
Source 2	Beta	Strontium 90
Source 3	Gamma + Beta	Cobalt 60

(*a*) (i) Describe an experiment to show which is the alpha emitting source.

Your description must include:

- equipment used

- measurements taken

- an explanation of the results. **3**

(ii) The radioactive material in Source 3 emits both beta and gamma radiations. Describe how the window of the casing could be modified so that the beta radiation is stopped. **1**

(*b*) Strontium 90 has a half life of 28 years. Calculate how many years it takes for the activity to decrease to 1/16th of its original value. **2**

(*c*) (i) A technician working with Source 1 receives an absorbed dose of $20 \, \mu\mathrm{Gy}$ of alpha particles. Calculate the total equivalent dose received by the technician. **2**

(ii) Describe two ways in which the technician could reduce his absorbed dose. **2**

(10)

[*END OF QUESTION PAPER*]

[BLANK PAGE]

[BLANK PAGE]

X069/201

NATIONAL	FRIDAY, 28 MAY	PHYSICS
QUALIFICATIONS	1.00 PM – 3.00 PM	INTERMEDIATE 2
2010		

Read Carefully

Reference may be made to the Physics Data Booklet

1   All questions should be attempted.

Section A (questions 1 to 20)

2   Check that the answer sheet is for Physics Intermediate 2 (Section A).

3   For this section of the examination you must use an **HB pencil** and, where necessary, an eraser.

4   Check that the answer sheet you have been given has **your name, date of birth, SCN** (Scottish Candidate Number) and **Centre Name** printed on it.

Do not change any of these details.

5   If any of this information is wrong, tell the Invigilator immediately.

6   If this information is correct, **print** your name and seat number in the boxes provided.

7   There is **only one correct** answer to each question.

8   Any rough working should be done on the question paper or the rough working sheet, **not** on your answer sheet.

9   At the end of the exam, put the **answer sheet for Section A inside the front cover of your answer book.**

10   Instructions as to how to record your answers to questions 1–20 are given on page three.

Section B (questions 21 to 30)

11   Answer the questions numbered 21 to 30 in the answer book provided.

12   **All answers must be written clearly and legibly in ink.**

13   Fill in the details on the front of the answer book.

14   Enter the question number clearly in the margin of the answer book beside each of your answers to questions 21 to 30.

15   Care should be taken to give an appropriate number of significant figures in the final answers to calculations.

DATA SHEET

Speed of light in materials

Material	Speed in m/s
Air	$3 \cdot 0 \times 10^8$
Carbon dioxide	$3 \cdot 0 \times 10^8$
Diamond	$1 \cdot 2 \times 10^8$
Glass	$2 \cdot 0 \times 10^8$
Glycerol	$2 \cdot 1 \times 10^8$
Water	$2 \cdot 3 \times 10^8$

Speed of sound in materials

Material	Speed in m/s
Aluminium	5200
Air	340
Bone	4100
Carbon dioxide	270
Glycerol	1900
Muscle	1600
Steel	5200
Tissue	1500
Water	1500

Gravitational field strengths

	Gravitational field strength on the surface in N/kg
Earth	10
Jupiter	26
Mars	4
Mercury	4
Moon	1·6
Neptune	12
Saturn	11
Sun	270
Venus	9

Specific heat capacity of materials

Material	Specific heat capacity in J/kg °C
Alcohol	2350
Aluminium	902
Copper	386
Glass	500
Ice	2100
Iron	480
Lead	128
Oil	2130
Water	4180

Specific latent heat of fusion of materials

Material	Specific latent heat of fusion in J/kg
Alcohol	$0 \cdot 99 \times 10^5$
Aluminium	$3 \cdot 95 \times 10^5$
Carbon dioxide	$1 \cdot 80 \times 10^5$
Copper	$2 \cdot 05 \times 10^5$
Iron	$2 \cdot 67 \times 10^5$
Lead	$0 \cdot 25 \times 10^5$
Water	$3 \cdot 34 \times 10^5$

Melting and boiling points of materials

Material	Melting point in °C	Boiling point in °C
Alcohol	−98	65
Aluminium	660	2470
Copper	1077	2567
Glycerol	18	290
Lead	328	1737
Iron	1537	2747

Specific latent heat of vaporisation of materials

Material	Specific latent heat of vaporisation in J/kg
Alcohol	$11 \cdot 2 \times 10^5$
Carbon dioxide	$3 \cdot 77 \times 10^5$
Glycerol	$8 \cdot 30 \times 10^5$
Turpentine	$2 \cdot 90 \times 10^5$
Water	$22 \cdot 6 \times 10^5$

Radiation weighting factors

Type of radiation	Radiation weighting factor
alpha	20
beta	1
fast neutrons	10
gamma	1
slow neutrons	3

SECTION A

For questions 1 to 20 in this section of the paper the answer to each question is either A, B, C, D or E. Decide what your answer is, then, using your pencil, put a horizontal line in the space provided—see the example below.

EXAMPLE

The energy unit measured by the electricity meter in your home is the

  A kilowatt-hour

  B ampere

  C watt

  D coulomb

  E volt.

The correct answer is **A**—kilowatt-hour. The answer **A** has been clearly marked in **pencil** with a horizontal line (see below).

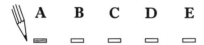

Changing an answer

If you decide to change your answer, carefully erase your first answer and, using your pencil, fill in the answer you want. The answer below has been changed to **E**.

A   B   C   D   E

[Turn over

SECTION A

Answer questions 1–20 on the answer sheet.

1. Which of the following is a scalar quantity?

   A   Force

   B   Acceleration

   C   Momentum

   D   Velocity

   (E)   Energy

2. A student investigates the speed of a trolley as it moves down a slope.

   The apparatus is set up as shown.

   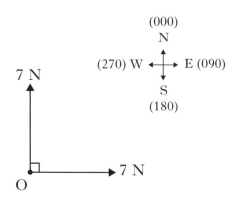

   The following measurements are recorded.

   distance from P to Q = 1·0 m
   length of card on trolley = 0·04 m
   time taken for trolley to travel from P to Q = 2·5 s
   time taken for card to pass through light gate = 0·05 s

   The speed at Q is

   A   0·002 m/s

   B   0·016 m/s

   C   0·40 m/s

   (D)   0·80 m/s

   E   20 m/s.

3. Two forces, each of 7 N, act on an object O.

   The forces act as shown.

   $$
   \begin{array}{c}
   (000) \\
   N \\
   (270)\ W \longleftrightarrow E\ (090) \\
   S \\
   (180)
   \end{array}
   $$

   The resultant of these two forces is

   A   7 N at a bearing of 135

   (B)   9·9 N at a bearing of 045

   C   9·9 N at a bearing of 135

   D   14 N at a bearing of 045

   E   14 N at a bearing of 135.

4 A package is released from a helicopter flying horizontally at a constant speed of 40 m/s.

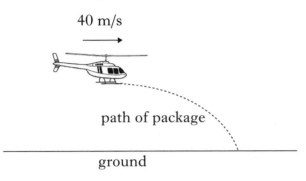

40 m/s

path of package

ground

The package takes 3·0 s to reach the ground.

The effects of air resistance can be ignored.

Which row in the table shows the horizontal speed and vertical speed of the package just before it hits the ground?

	Horizontal speed (m/s)	Vertical speed (m/s)
A	0	30
B	30	30
C	30	40
D	40	30
E	40	40

$v \propto \dfrac{d}{t}$

5. 100 g of a solid is heated by a 50 W heater. The graph of temperature of the substance against time is shown.

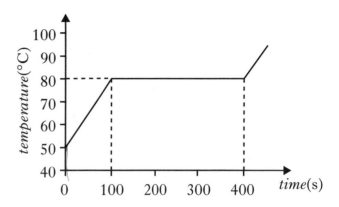

The specific latent heat of fusion of the substance is

A $1·3 \times 10^3$ J/kg

B $1·5 \times 10^3$ J/kg

C $3·0 \times 10^3$ J/kg

D $1·5 \times 10^5$ J/kg

E $1·9 \times 10^5$ J/kg.

[Turn over

$P = E \div t$

$E = \dfrac{P}{t}$

$= \dfrac{50}{}$

$= 0.5 \, \text{J}$

ET $E = Pt$

$= 50 \times 60$

$=$

$E = mL$

$0.5 = mL$

$\dfrac{E}{mL}$

$L = \dfrac{E}{m}$

$= \dfrac{0.5}{0.1}$

$=$

6. A crate of mass 200 kg is pushed a distance of 20 m across a level floor.

   The crate is pushed with a force of 150 N.

   The force of friction acting on the crate is 50 N.

   The work done in pushing the crate across the floor is

   $E_W = Fd$
   $= 150 \times 20$
   $=$

   A   1000 J

   Ⓑ   2000 J

   C   3000 J

   D   4000 J

   E   20 000 J.

7. A student makes the following statements about electrical circuits.

   I The sum of the potential differences across components connected in series is equal to the supply voltage.

   II The sum of the currents in parallel branches is equal to the current drawn from the supply.

   III The potential difference across components connected in parallel is the same for each component.

   Which of the statements is/are correct?

   A   I only

   B   III only

   C̶   I and II only

   D   II and III only

   Ⓔ   I, II and III

8. Three resistors are connected as shown

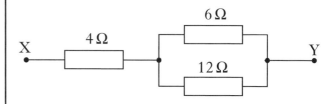

   The total resistance between X and Y is

   A   2 Ω

   B   4 Ω

   Ⓒ   8 Ω

   D   13 Ω

   E   22 Ω.

9. The resistance of a wire is 6 Ω.

   The current in the wire is 2 A.

   The power developed in the wire is

   A   3 W

   B   12 W

   C   18 W

   Ⓓ   24 W

   E   72 W.

   $P = VI$

   $P = \dfrac{V^2}{R}$ $P = I^2 R$
   $=$

10. The voltage of the mains supply in the UK is 230 V a.c.

Which row in the table shows the peak voltage and frequency of the mains supply in the UK?

	peak voltage (V)	frequency (Hz)
A	175	50
B	175	60
C	230	50
D	325	50
E	325	60

11. The diagram shows a model bicycle dynamo.

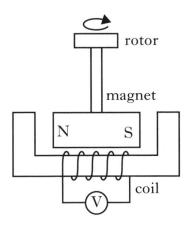

When the rotor is turned the magnet rotates, inducing a voltage in the coil. The induced voltage can be <u>decreased</u> by

A   increasing the number of turns on the coil

B   decreasing the number of turns on the coil

C   using a stronger magnet

D   turning the rotor faster

E   reversing the direction of rotation of the magnet.

12. The graph below shows how the input voltage V_1 to an amplifier varies with time t.

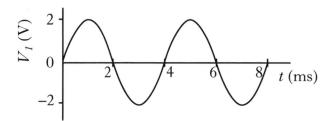

The amplifier has a voltage gain of 10.

Which graph shows how the output voltage V_0 of the amplifier varies with time t?

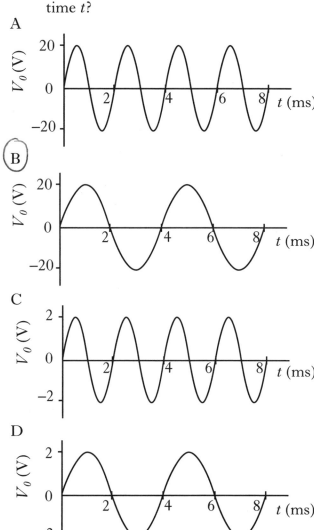

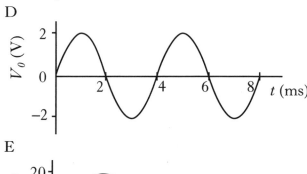

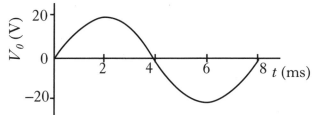

13. The diagram gives information about a wave.

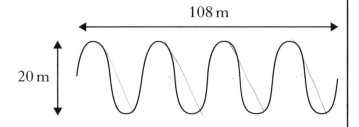

108 m

20 m

The time taken for the waves to travel 108 m is 0·5 s.

A student makes the following statements about the waves.

I The wavelength of the waves is 27 m.

II The amplitude of the waves is 20 m.

III The frequency of the waves is 8 Hz.

Which of the statements is/are correct?

A I only

B II only

C I and III only

D II and III only

E I, II and III

14. The diagram shows members of the electromagnetic spectrum in order of increasing wavelength.

Gamma rays	P	Ultraviolet radiation	Q	Infrared radiation	R	TV & radio waves

⸻ increasing wavelength ⟶

Which row in the table identifies the radiations represented by the letters P, Q and R?

	P	Q	R
A	X-rays	visible light	microwaves
B	X-rays	microwaves	visible light
C	microwaves	visible light	X-rays
D	visible light	microwaves	X-rays
E	visible light	X-rays	microwaves

15. An object is placed in front of a converging lens as shown.

The position of the image formed by the lens is also shown.

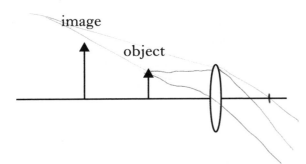

The focal length of the lens is 100 mm.

The distance between the lens and the object is

A   50 mm

B   100 mm

C   150 mm

D   200 mm

E   250 mm.

16. A converging lens has a focal length of 50 mm.

The power of the lens is

A   +0·02 D

B   +0·2 D

C   −0·2 D

D   +20 D

E   −20 D.

$$P = \frac{1}{50}$$

17. A student makes the following statements about a carbon atom.

  I The atom is made up only of protons and neutrons.

  II The nucleus of the atom contains protons, neutrons and electrons.

  III The nucleus of the atom contains only protons and neutrons.

Which of the statements is/are correct?

A   I only

B   II only

C   III only

D   I and II only

E   I and III only

18. Human tissue can be damaged by exposure to radiation.

On which of the following factors does the risk of biological harm depend?

  I The absorbed dose.

  II The type of radiation.

  III The body organs or tissue exposed.

A   I only

B   I and II only

C   II only

D   II and III only

E   I, II and III

[Turn over

19. Information about a radioactive source is given in Table 1.

Table 1

Activity	Energy absorbed per kilogram of tissue	Radiation weighting factor
500 MBq	0·2 µJ	10

Which row in Table 2 gives the correct information for the radioactive source?

Table 2 $H = D \times R =$

	Absorbed Dose	Equivalent Dose
A	0·2 µGy	2 µSv
B	500 MGy	10 Sv
C	10 Gy	0·2 µSv
D	20 µGy	50 MSv
E	2 µGy	0·2 µSv

20. In a nuclear reactor a chain reaction releases energy from nuclei.

Which of the following statements describes the beginning of a chain reaction?

A   An electron splits a nucleus releasing more electrons.

B   An electron splits a nucleus releasing protons.

C   A proton splits a nucleus releasing more protons.

D   A neutron splits a nucleus releasing electrons.

E   A neutron splits a nucleus releasing more neutrons.

Candidates are reminded that the answer sheet for Section A MUST be placed INSIDE the front cover of the answer book.

SECTION B   *Marks*

Write your answers to questions 21–30 in the answer book.

All answers must be written clearly and legibly in ink.

21. A balloon of mass 400 kg rises vertically from the ground.

The graph shows how the vertical speed of the balloon changes during the first 100 s of its upward flight.

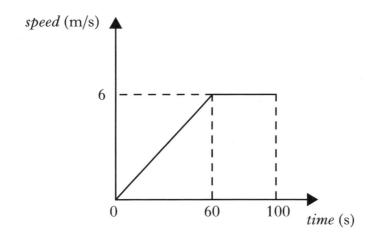

(a) Calculate the acceleration of the balloon during the first 60 s.   2

(b) Calculate the distance travelled by the balloon in 100 s.   2

(c) Calculate the average speed of the balloon during the first 100 s.   2

(d) Calculate the weight of the balloon.   2

(e) Calculate the total upward force acting on the balloon during the first 60 s of its flight.   3

(11)

Marks

22. Inside a storm cloud water droplets move around and collide with each other.

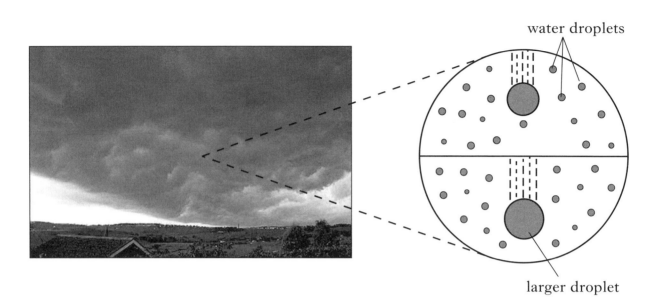

water droplets

larger droplet

(a) A water droplet of mass 2·0 g moving at a speed of 4·0 m/s collides with a stationary water droplet of mass 1·2 g. The two droplets join together to form a larger droplet.

Calculate the speed of this larger droplet after the collision.   2

(b) Another water droplet within the cloud is falling with a constant speed. Draw a diagram showing the forces acting on this droplet.

Name these forces and show their directions.   2

(c) The motion of water droplets in the cloud causes flashes of lightning. One lightning flash transfers 1650 C of charge in 0·15 s.

Calculate the electric current produced by this flash.   2

(d) Why does an observer, standing 3 km from a thunder cloud, see a lightning flash before he hears the thunder?   1

(7)

Marks

23. On the planet Mercury the surface temperature at night is −173 °C. The surface temperature during the day is 307 °C. A rock lying on the surface of the planet has a mass of 60 kg.

(a) The rock absorbs $2·59 \times 10^7$ J of heat energy from the Sun during the day.

Calculate the specific heat capacity of the rock.   2

(b) Heat is released at a steady rate of 1440 J/s at night.

Calculate the time taken for the rock to release $2·59 \times 10^7$ J of heat.   2

(c) Energy from these rocks could be used to heat a base on the surface of Mercury.

How many 60 kg rocks would be needed to supply a 288 kW heating system?   2

(d) Using information from the data sheet, would it be **easier**, **the same** or **more difficult** to lift rocks on Mercury compared to Earth?

You **must** explain your answer.   2

(8)

[Turn over

Marks

24.   A student sets up the following circuit.

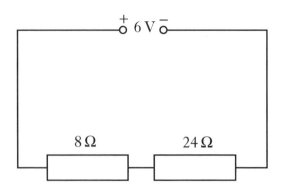

(a) Calculate the current in the 8 Ω resistor.   **3**

(b) Calculate the voltage across the 8 Ω resistor.   **2**

(c) The 24 Ω resistor is replaced by one of **greater** resistance. How will this affect the voltage across the 8 Ω resistor?

Explain your answer.   **2**

(7)

Marks

25. In a lab experiment a technician builds a transformer and uses electrical meters to take a number of measurements, as shown in the diagram.

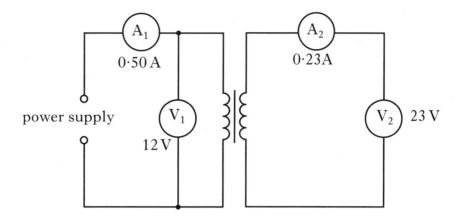

(*a*) The technician has a choice of an a.c. or a d.c. power supply. Which power supply should be used?

Explain your answer. 2

(*b*) Calculate the electrical power in the primary circuit of the transformer. 2

(*c*) Calculate the electrical power in the secondary circuit of the transformer. 1

(*d*) Calculate the percentage efficiency of the transformer. 2

(*e*) Another experiment uses a different transformer. It is 100% efficient. The primary coil has 1500 turns and the secondary coil contains 3000 turns.

Calculate the secondary voltage when the primary voltage is 12 V. 2

(9)

[Turn over

Marks

26. Water in a fish tank has to be maintained at a constant temperature. Part of the electronic circuit which controls the temperature is shown.

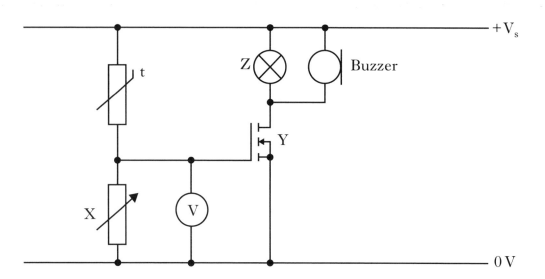

(a) Name components Y and Z.   **2**

(b) What happens to the resistance of the thermistor as the temperature increases?   **1**

(c) When the voltmeter reading reaches 1·8 V component Y switches on. Explain how the circuit operates when the temperature rises.   **2**

(d) Why is a variable resistor chosen for component X rather than a fixed value resistor?   **1**

(6)

Marks

27. At the kick-off in a football match, during the World Cup Finals, the referee blows his whistle. The whistle produces sound waves.

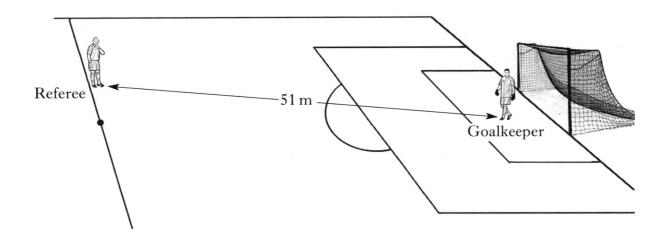

(a) Using information from the diagram and the data sheet, calculate the time taken for the sound waves to reach the goalkeeper.   2

(b)   (i) Are sound waves transverse or longitudinal waves?   1

    (ii) Describe **two** differences between transverse and longitudinal waves.   2

  (iii) What is transferred by waves?   1

(c)   (i) Floodlights in the stadium are switched on. Each lamp has a power rating of 2·40 kW. The operating voltage is 315 V.

    Calculate the resistance of a lamp.   2

    (ii) The floodlights consist of 20 lamps connected in parallel.

    State **two** reasons why the lamps are connected in parallel.   2

            (10)

[Turn over

Marks

28. A satellite sends microwaves to a ground station on Earth.

   (*a*) The microwaves have a wavelength of 60 mm.

      (i) Calculate the frequency of the waves.　　2

      (ii) Determine the period of the waves.　　2

   (*b*) The satellite sends radio waves along with the microwaves to the ground station. Will the radio waves be received by the ground station **before**, **after** or **at the same time** as the microwaves?

      Explain your answer.　　2

   (*c*) When the microwaves reach the ground station they are received by a curved reflector.

      Explain why a curved reflector is used.

      Your answer may include a diagram.　　2

        (8)

Marks

29. In 1908 Ernest Rutherford conducted a series of experiments involving alpha particles.

(a) State what is meant by an alpha particle. 1

(b) Alpha particles produce a greater ionisation density than beta particles or gamma rays. What is meant by the term *ionisation*? 1

(c) A radioactive source emits alpha particles and has a half-life of 2·5 hours. The source has an initial activity of 4·8 kBq.

Calculate the time taken for its activity to decrease to 300 Bq. 2

(d) Calculate the number of decays in the sample in two minutes, when the activity of the source is 1·2 kBq. 2

(e) Some sources emit alpha particles and are stored in lead cases despite the fact that alpha particles cannot penetrate paper. Suggest a possible reason for storing these sources using this method. 1

(7)

[Turn over for Question 30 on *Page twenty*

Marks

30. Many countries use nuclear reactors to produce energy. A diagram of the core of a nuclear reactor is shown.

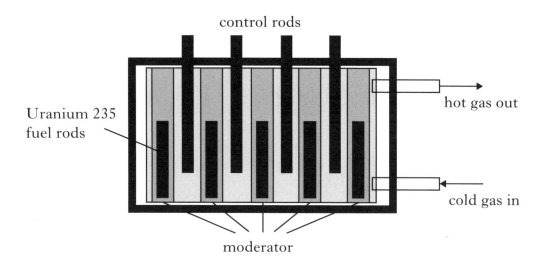

(*a*) State the purpose of:

(i) the moderator;   **1**

(ii) the control rods.   **1**

(*b*) One nuclear fission reaction produces $2 \cdot 9 \times 10^{-11}$ J of energy. The power output of the reactor is $1 \cdot 4$ GW. How many fission reactions are produced in one hour?   **3**

(*c*) State **one advantage** and **one disadvantage** of using nuclear power for the generation of electricity.   **2**

(7)

[END OF QUESTION PAPER]

[BLANK PAGE]

[BLANK PAGE]

[BLANK PAGE]

[BLANK PAGE]

[BLANK PAGE]

[BLANK PAGE]

[BLANK PAGE]

Acknowledgements

Permission has been sought from all relevant copyright holders and Bright Red Publishing is grateful for the use of the following:
Picture of an exercise bike taken from www.multisportfitness.com © MultiSports, Inc (2008 page 11).